PHILOSOPHICAL LOGIC AND LOGICAL PHILOSOPHY

SYNTHESE LIBRARY

STUDIES IN EPISTEMOLOGY,

LOGIC, METHODOLOGY, AND PHILOSOPHY OF SCIENCE

VOLUME 257

VLADIMIR ALEXANDROVICH SMIRNOV
(March 2, 1931 - February 12, 1996)

PHILOSOPHICAL LOGIC AND LOGICAL PHILOSOPHY

Essays in Honour of Vladimir A. Smirnov

Edited by

PETER I. BYSTROV

Institute of Philosophy, Russian Academy of Sciences,
Moscow, Russia

and

VADIM N. SADOVSKY

Institute for Systems Analysis, Russian Academy of Sciences,
Moscow, Russia

KLUWER ACADEMIC PUBLISHERS

DORDRECHT / BOSTON / LONDON

A C.I.P. Catalogue record for this book is available from the Library of Congress

ISBN 0-7923-4270-4

Published by Kluwer Academic Publishers,
P.O. Box 17, 3300 AA Dordrecht, The Netherlands.

Kluwer Academic Publishers incorporates
the publishing programmes of
D. Reidel, Martinus Nijhoff, Dr W. Junk and MTP Press.

Sold and distributed in the U.S.A. and Canada
by Kluwer Academic Publishers,
101 Philip Drive, Norwell, MA 02061, U.S.A.

In all other countries, sold and distributed
by Kluwer Academic Publishers Group,
P.O. Box 322, 3300 AH Dordrecht, The Netherlands.

Printed on acid-free paper

Printed in the Netherlands

Table of Contents

Part V. Analysis of Logical Proofs

Preface

Vladimir Aleksandrovich Smirnov was born on March 2, 1931. He graduated from Moscow State University in 1954. From 1957 till 1961 he was a lecturer in philosophy and logic at the Tomsk University. Since 1961 his scientific activity continued in Moscow at the Institute of Philosophy of Academy of Sciences of the USSR. From 1970 and till the last days of his life V. A. Smirnov was lecturer and then Professor at the Chair of Logic at Moscow State University.

V. A. Smirnov played an important role at the Institute of Philosophy of Russian Academy of Sciences being the Head of Department of Epistemology, Logic and Philosophy of Science and Technology, and the Head of Section of Logic. Last years he was the leader of the Centre of Logical Investigations of Russsian Academy of Sciences. In 1990–91 he founded a new non-goverment Institute of Logic, Cognitive Sciences and Development of Personality for performing research, teaching, editorial and organization activity in the field of humanities. At the Department of Philosophy of Moscow State University and at the Institute of Philosophy V. A. Smirnov and his close colleagues have founded a Russian logical school which brought up many talented researchers who work at several scientific centres in various countries.

Undoubtedly, V. A. Smirnov was one of the most prominent figures in logic and methodology of science in the former USSR and Russia. He published two books – "Formal Deduction and Logical Calculi" and "Logical Methods of Analysis of Scientific Theories" which have become classics for Russian logicians and are well-known in several countries. Interesting ideas and results are contained in his 150 published articles and numerous manuscripts. He worked fruitfully in many fields of modern logic, especially in non-classical logics and its applications to various scientific disciplines, and obtained a number of original results pertaining to different forms of logical deduction, interconnections between syllogistics, predicate logic and algebra, logical structure of scientific theories, fundamental properties of formal logical systems, etc. He proposed non-standard solutions to various philosophical and methodological problems concerned with the interconnection between empirical and theoretical levels of scientific knowledge, the status and functions of ideal objects, and features of rational creativity. Recently he was successfully elaborating the theoretical and methodological foundations of intellectual computer-based systems.

What is to be especially mentioned is V. A. Smirnov's organizational activity. On the one hand it promoted increasing influence and impact of logic and rational philosophical methods in the scientific community of the former Soviet Union. He was one of the main organizers of All-Union Conferences in Logic, Methodology and Philosophy of Sciences from the early 1960s. During the last fifteen years he was the leader of the Scientific Seminar in Logic at the Institute of Philosophy where the newest logical results were presented and discussed by the leading logicians from Russia and many other countries. These were precisely the Conferences and Seminars that gave a good impetus for scientific collaboration of logicians, philosophers, mathematicians, scholars in linguistics and computer sciences. On the other hand, his constructive efforts were very important for carrying out the fruitful scientific contacts between logicians from the former USSR and other countries. He was one of the organizers of the Soviet-Finnish, Soviet-Polish, Soviet-Japan Logical Conferences and Soviet-American Joint Scientific Seminar in Philosophy and Computer Sciences. He was the active participant in many World and International meetings. In particular, he took part in every International Congresses of Logic, Methodology and Philosophy of Science, from the third to the tenth Congress.

Initially we intended to publish this Collection in honour of Vladimir Aleksandrovich Smirnov on occasion of his birthday. However, serious illness attacked him suddenly. He knew about it, he fought against it and worked till the last minutes of his life. He died on February 12, 1996.

Vladimir Smirnov is still alive in the hearts and memories of his friends, colleagues and pupils from many countries. The authors of this Collection would like to pay homage to this remarkable man as a scholar, colleague, teacher and friend. All the essays of this Collection touch upon topics in the sphere of logic, methodology and philosophy of science in which V.A.Smirnov has taken an interest. Now they are published in his memory.

Peter Bystrov
Jaakko Hintikka
Vadim Sadovsky

PART I

Epistemology and Philosophy of Science

1

Understanding Scientific Change

Veikko Rantala

1 Introduction

In his article 'Commensurability, Comparability, Communicability', Thomas Kuhn distinguishes between translation and interpretation and indicates that some of his critics incorrectly equate the two notions.[1] Translation involves an interpretive component, but it is not the same as interpretation. Interpreting a culture, language, scientific theory or paradigm may involve learning a new language or possibly an earlier version of the interpreter's own language. According to Kuhn, however, learning a new language is not the same as translating it into the interpreter's own language. Interpreting does not imply translating. On the other hand, incommensurability in this article means the lack of an adequate translation: there is no language into which the two languages or theories in question can be translated without residue.

From this article onwards, Kuhn's notion of translation is very strict: a translation of a text into another language is a systematic replacement of the words and strings of words of the text by words and strings of words of the other language so as to produce an equivalent text.[2] Here equivalence means that the two texts have the same extension or the same intension.[3] In the translation, expressions of the former language must correspond to coreferential expressions of the latter. Furthermore, the structure of the world determined by the lexical structure of the former language is preserved.

It seems that this notion of translation (which, however, is in conformity with the standard view of translation) is so strong that it excludes almost any nontrivial, perfect translation (thus commensurability), and he comes very close to saying it himself as he argues, for example, that different languages impose different structures on the world. In particular, his notion does not only make two theories

[1] [7].

[2] Cf. e.g. [6], where his notion of translation is very liberal. For discussions of this early notion, see also [8, 20].

[3] But Kuhn seems to vacillate between these two alternatives.

P. I. Bystrov and V. N. Sadovsky (eds.), Philosophical Logic and Logical Philosophy, 3–15.

belonging to different paradigms mutually untranslatable but also theories in the same paradigm. However, rather than dwelling on these much debated negative arguments stated by Kuhn about intertheoretic relations, I shall in this paper consider the question of whether there exists a notion of translation which would be more useful when one attempts to understand scientific change and the role of such notions as interpretation, in a sense discussed by Kuhn, explanation, and reduction.[4] Translation is necessary for reduction which in turn may yield explanation. If one theory, say T, is to be reduced to another, T', the reduction usually presupposes a translation, of some kind, of the language of T into the language of T'. On the other hand, the reduction may make it possible to explain T by means of T', in some sense of 'explanation'. The explanation, in turn, increases our understanding of the respective scientific change, that is, of the change from T to T', for it is a way of looking at T from the perspective of T', or, rather, from the perspective of the paradigm to which T' belongs.

In this paper, I shall mainly consider the understanding of scientific change in the above sense, but I shall also touch on the question from the point of view of the reduced theory T. Historically, scientists are often familiar with the reduced theory in the first place, that is, T exists first and T' is discovered later, as, e.g., in a case where T' supplants T. Then one has to look at T' from the perspective of T and its paradigm in order to understand the nature of the change in its historical situation, in its 'context of discovery'. Both perspectives are needed for the understanding of the nature of scientific change—and for the understanding of philosophers' disagreement about it.

2 Explanation

If one theory or law, or lawlike sentence, is to be explained by another, various social, cognitive, and intentional presuppositions must be satisfied. The explanation is pragmatically constrained. Therefore, any feasible model of intertheoretic explanation has to give an account of relevant pragmatic constraints.[5] The deductive-nomological model of explanation was designed to represent ideally objective explanation, where the role of pragmatic constraints is minimised. This ideal has been best realised in some noncontroversial applications of the model. Such easy applications are rare in the history of science, however. On the other hand, the more controversial its applicability is, the more prominent seems to be the role of pragmatic constraints, and the less warranted it is to talk about explaining theories by means of the D-N model. Consider, for instance, the notion of approximate deductive explanation (involving no counterfactual special assumptions) as applied to a scientific change where a theory is superseded by a new one. It is customary to say that since the new theory can be approximated by means of the superseded one,

[4] Thus I shall further elaborate some themes previously studied by David Pearce and myself.

[5] For a more thorough discussion of pragmatic aspects of explanation see, e.g., [25, 27]. [26] employs speech act theory and Hintikka's [5] question theory for studying pragmatic features of explanation.

this does not only explain why the latter is approximately true when it is, but it also explains why scientists used to hold such a 'false' theory, and why it is still successful as a calculating device. However, if the D-N model is applied here, the role of extensional logic is emphasised and the model, as it stands, does not contain any account of the behaviour of scientists or of any other pragmatic features.

There have been attempts to apply the notion of deductive explanation, in some form or another, to the most controversial cases of scientific change, that is, cases which are said to represent scientific revolution, incommensurability, limiting correspondence, and so forth, but these applications have appeared extremely problematic. It is above all the extensional, paradigm-independent character of the D-N model which has been problematic in such applications, since, as was pointed out by Kuhn [6] and others, such scientific changes are socially loaded: they cannot be described without giving an account of paradigmatic factors. It follows that no model of intertheoretic explanation is adequately applicable to revolutionary change unless relevant pragmatic constraints are embedded in it, and part of them must be there explicitly.

In an earlier work, I have proposed a model which would apply to such cases of revolutionary scientific change. They are changes which can be represented in terms of 'counterfactual reduction'; where, roughly speaking, a supplanted law is obtainable from a supplanting law by means of limiting or other counterfactual special conditions.[6] The model has a clear-cut logical structure and, on the other hand, it takes some pragmatic constraints into account by looking into the supplanted law from the perspective of the supplanting paradigm. The latter feature yields Kuhnian relativity, the former objectivity in the sense of something like D-N explanation. The model is obtained, roughly speaking, by putting together

1. a general notion of intertheoretic reduction as it has been developed by David Pearce and myself,
2. Clark Glymour's idea of counterfactual explanation,[7] and
3. Nelson Goodman's and David Lewis's approaches to counterfactuals.[8]

I shall first give a brief overview of the model. In what follows, I shall refer to scientific laws as axioms of theories; a law is something accepted as a law of some theory. It will be assumed that the laws to be considered can be reconstructed as sentences of appropriate logics. We may assume that a law sentence has models which are abstract, model-theoretic representations of the phenomena or events, actual or merely possible, described by the law. We shall also talk about possible worlds or situations here—and presuppose that these notions are properly understood as heuristic devices.

---

[6] See [18, 20]; [22].

[7] [3].

[8] [4, 9].

Let A and A' be laws of theories T and T', respectively, and let L and L' be logics in which A and A' are formulated.[9] Assume, for simplicity, that A and A' are the only axioms of T and T'. Then a *correspondence* of T to T' is a pair of mappings $\langle F, I\rangle$ such that F assigns models of A to specified models of A' and I is a translation of the language of L into the language of L'. The two mappings have to satisfy the following crucial condition: for all sentences S of L and for all models M of A' in the domain of F:

1. $F(M) \models_L S \Leftrightarrow M \models_{L'} I(S)$,

where $I(S)$ is the translation of S under I.[10]

The two mappings yield an informative conceptual relationship between the two theories only if they meet appropriate logical and pragmatic conditions. On some general assumptions concerning F, L and L', an argument of the following form can be inferred:

2. $A', C_0, \ldots, C_n / I(A)$,

where'/' indicates the logical consequence, or (if L' is complete) the deduction in L', and $C_0, \ldots, C_n$ are special assumptions, the nature of which will be discussed later.

If some of the special assumptions $C_0, \ldots, C_n$ are counterfactual (in a sense to be specified in each case separately), the correspondence $\langle F, I\rangle$ will be called a *counterfactual correspondence*. Let there be just one counterfactual assumption C, i.e., let C be the single special assumption with respect to which the correspondence is counterfactual. If the mappings F and I involve concepts of nonstandard analysis, and C is an appropriate limit condition, the nature of which depends on context,[11] $\langle F, I\rangle$ is called a *limiting case correspondence*. Then $I(A)$ is usually read as follows:

3. A almost holds,

where *almost* means infinitesimal accuracy, that is, that A holds for all practical purposes. The special assumption C may be considered possible in a logical sense, though it does not hold in any actual situations. If (2) refers to a limiting case correspondence, it implies that A almost holds in the situations where A' plus the special conditions hold. It does not follow that A would almost hold in actual situations.[12]

[9] As I pointed out in [16], the generalised notion of logic developed in abstract logic (general model theory) opens new methodological possibilities for metascientific study. There are in principle no restrictions as to the choice of logics L And L'; what logic is chosen in a given application is determined by pragmatic and purely logical reasons. See also [10, 12, 11].

[10] For some qualifications and generalisations, see my [21].

[11] For example, in the case of the relationship between classical particle mechanics and relativistic particle mechanics, C is a sentence formalising the condition 'the velocity of light is infinite'.

[12] At least some cases of scientific change which are usually classified as instances of limiting case correspondence can be reconstructed in this framework; see Pearce and Rantala [13, 14], Rantala and Vadén [22].

We have to ask now whether counterfactual correspondence is of any explanatory importance.[13] Let us first consider the import of the argument (2). If it satisfies appropriate formal and pragmatic conditions of adequacy, we might perhaps say that it provides a D-N explanation of $I(A)$. But what would then be a relevant question to which this explanation would provide an answer? Assume for the sake of argument, that we have here a limiting case correspondence. Then we cannot simply ask: '*Why* does A almost hold?' since A may not almost hold in any actual situations; and even if we could, (2) would not provide any adequate answer because one of the special assumptions is counterfactual. Rather, a relevant why-question would be: '*Why* does A almost hold in the (possible) situations where it does so?'. But since one of the special assumptions is contrary-to-fact, of what explanatory import would it be to correctly answer that A almost holds in those situations *because* A' holds generally and the special assumptions hold in the same situations and they logically imply $I(A)$? The only kind of explanatory import one can imagine the argument to yield is conceptual in the following sense: by studying the special assumptions and the translation $I(A)$, one learns something about the conceptual relationship of the two laws. In earlier papers, I have pointed out that if the correspondence relation $\langle F, I \rangle$ itself is examined, in addition to its consequence (2), one may learn more about the relationship. This fact is not just logical, but can be seen from studies of actual scientific theories.[14] Thus, it seems that even though we were justified in claiming that (2) provides an instance of D-N explanation—I do not think anymore that we are justified—its explanatory role *as* a D-N explanation would be very weak.[15] We have, as it were, to shift the emphasis from explanation to understanding. To see this, I shall next elaborate Glymour's [3] proposal concerning the explanatory import of counterfactual conditions.[16]

Glymour suggests that instead of why-questions we have to ask different questions when counterfactual special assumptions are involved. One must ask, for instance: 'Under what conditions would the supplanted law hold?'. The law is then explained by:

4a. showing under what conditions it would hold;

4b. contrasting those conditions with the conditions which actually obtain.

Glymour has here a deductive picture of explanation in mind, however. His idea is, roughly, that an inference where some of the premises (or special assumptions, as he calls them) are counterfactual and some connect the languages of the

[13] Let us recall at this point that what is here taken as translation is too general to be called 'translation' by Kuhn [7]: condition (1) does not in general meet his criteria of adequacy. Balzer *et al.* [1] also consider it too weak to represent a translation proper since it does not preserve meaning. It follows that the existence of a translation in the present sense does not imply commensurability in the sense of Kuhn [7]. On the other hand, it has been argued by Pearce [11] that Kuhn's notion of commensurability can be challenged and replaced by weaker notions which admit of the cognitive comparability of theories in a sufficiently strict sense.

[14] This can be seen by carefully investigating the case studies in the articles referred to in footnote 12.

[15] This view is in fact somewhat contrary to what was maintained in Pearce and Rantala [15].

[16] For a more detailed exposition see my [21].

two laws, does the job (4a). (4b) is accomplished by showing that without the counterfactual premises the supplanted law is not in general entailed by the supplanting law.

Let us consider this idea more closely; I restrict myself here to (4a).[17] (4b) yields an answer of the form 'Since such-and-such conditions do not hold' to the question 'Why the law does not hold?', and that may increase our understanding of the law, but what its explanatory import is is not very clear. Assuming, as before, that C is the only counterfactual premise, (4a) amounts to showing that a conditional of the following form is true:

5. If it were the case that C, then it would be the case that A,

where A is the supplanted law. Hence, Glymour seems to suggest that by means of a relevant deduction one can establish (5). If this is correct, it means that Bonevac [2] is not quite accurate when arguing that in the formal sense the counterfactual premises are the only features that distinguish Glymour's analysis from a straightforwardly deductive account of explanation.

To embed Glymour's idea in the framework of counterfactual correspondence, the following small but essential correction must be made in his model: instead of including possible translations in the auxiliary premises, the conclusion A is to be replaced by its translation $I(A)$. Thus we say, for example, that 'it is almost the case that A'. Instead of (5), let us, therefore, consider the following conditional:

6. If it were the case that C, then it would be the case that $I(A)$,

Consider the argument (2). Let C be C_0 in (2); rewrite (2) into the form:

7. $A', C_1, \ldots, C_n, C / I(A)$

We may interpret Glymour as essentially claiming that the argument (7) makes the conditional (6) true. Whether (7) makes (6) true is not obvious, however, as we know from theories of counterfactual conditionals. It follows from these theories that the relationship between (6) and (7) must be studied more closely and that only on some pragmatic conditions (7) would establish the truth of (6). Such conditions may yield an explanation.

When does (7) make (6) true? The semantics of counterfactuals is intensional, whence the truth is often considered in the context of 'possible worlds'—or situations, events, occasions, or however one likes to call them. According to Lewis [9]—which naturally is not the only approach to counterfactuals[18]—(6) is true at a world w if there exists a valid argument of the form (7) *backing* (6) at w. We can as well say that if an argument is backing (6), it *makes* (6) *true* at w. (7) makes (6) true at w if each of the premises $A', C_1, \ldots, C_n$ is *cotenable* with the antecedent C at w. However, we are not in fact asking here whether there exists an argument

[17] For a somewhat more precise study, see Rantala [18].

[18] Any intensional semantics would help us to see the import of contextual and intentional aspects.

of the form (7) that makes (6) true at w, or whether (6) is true on some other truth condition, but whether it is true at w because (7) itself makes (6) true at w.

Therefore, we must ultimately study whether the following fact holds:

8. The premises $A', C_1, \ldots, C_n$ are cotenable with C at w,

where w is appropriately chosen. Whether (8) holds is, according to Lewis's theory, relative to the notions of *possibility* at w and *similarity* of possible worlds; that is, the answer depends on how these notions are interpreted and how w is chosen in this context. Assume first that C is possible at w. Then a premiss P is cotenable with C at w if

(i) P is necessary at w (this is a trivial case) or

(ii a) P and C are jointly possible, i.e., both true at some world u which is accessible from w, and

(ii b) P is true at every world which is at least as similar to w as u is.

C is assumed to be contrary-to-fact with respect to w, that is false at w. Thus, intuitively speaking, condition (ii) can be construed as saying that

(a) P is 'strongly enough possible', or, perhaps, 'sufficiently necessary', at w to be compatible with C no matter how 'implausible' C is (if only possible) when looked at from a point of view of w, and

(b) P is true at every world which resembles w (in relevant respects) 'as much as possible'.

Notice, furthermore, that if C is impossible at w, then the above definitions imply that the counterfactual (6) is true at w. It is true in a trivial sense, however, since the negation of C is then cotenable with C and $I(A)$ can be inferred from C and the negation; thus a backing argument can be found, but the argument is trivial. In such a case, the specific argument (7) makes (6) true at w only if the premises $A', C_1, \ldots, C_n$ are necessary at w.

Thus it appears in the light of David Lewis's theory that, contrary to what Glymour seems to say, an argument like (7) does not automatically make the respective conditional (6) true. It has been pointed out in [18] that the following more complex model of counterfactual explanation emerges when Glymour's model is improved by combining the generalised notion of counterfactual correspondence and Lewis's semantics: A *counterfactual explanation* of A from A', relative to w, consists of the following steps:

(i) Constructing a counterfactual correspondence $\langle F, I\rangle$ of A to A' (with respect to a special assumption C);

(ii) Deriving an argument (7) from $\langle F, I\rangle$;

(iii) Showing that the argument (7) makes the counterfactual (6) true at w.

At least the following pragmatic constraints are embedded in this model, either explicitly or implicitly:

(a) Explanation is to be performed from a point of view of the world w;

(b) $\langle F, I \rangle$ must be informative and constructible from what is known about the laws, paradigms, and logics involved;

(c) C must be possible at w;

(d) $A', C_1, \ldots, C_n$ must be cotenable with C, hence also true at w.

The constraints (a), (c)–(d) mean that the explanation is only good relative to a properly chosen context w. (c)–(d) are relative to the individual or scientific community giving the explanation, and to the adopted scientific principles. If the explainer holds the explaining law A' rather than A—which seems likely if A' has superseded A—then those scientific principles are determined by A' and the respective paradigmatic assumptions, rather than by A and its paradigm. In any case, there are no absolute notions of possibility, cotenability, and truth which would be available here. Constraint (b), which is not explicit in the model, and which is more or less vague, is intended to imply that not very much is explained by the mere knowledge of the existence of the correspondence $\langle F, I \rangle$.

In an earlier work, I have studied a couple of examples which illuminate the role of the above components of counterfactual explanation and the role of the pragmatic constraints.[19] I shall not repeat the examples here, but instead touch some conclusions which can be derived from them. The most striking conclusion, already hinted at above, seems to be that an explanation of the above kind of a superseded law can hardly be entirely 'objective'. The pragmatic constraints involved show that one has to look at the matter from the point of view of the new paradigm. This is what Kuhn [6] argues.[20] There may even be individual disagreement, concerning, for instance, the question of what counterfactual conditions can be considered possible. For instance, when the limiting correspondence between classical particle mechanics and relativistic particle mechanics is studied, one has to consider the question whether it is in some sense possible that the velocity of light is infinite. It is likely that there are philosophers and physicists who hold that this assumption is so much contrary-to-fact that it is not even possible and others who think that it is possible. There are elements here which leave much room for interpretation.[21]

[19] [18].

[20] See especially pp. 102–103.

[21] There is also methodological relativity here: many results are of course dependent on the methodological and logical tools one is employing.

3 Understanding Scientific Change

We have seen that the notion of deductive explanation is not appropriate for the most important, actual cases of scientific change. It does not provide a model of intertheoretic explanation in cases where the theories in question are, in the Kuhnian sense, incommensurable. In this paper, I have discussed a model of explanation which in a more general way would be applicable to theories which are in the relation of counterfactual correspondence to each other.[22]

This model cannot be defined in a purely extensional framework—not even Glymour's portion of the model, contrary to what Glymour [3] and Bonevac [2] seem to maintain—and its intensional features and pragmatic constraints bring in relative and subjective elements. Whether the counterfactual model can be successfully employed in a given case is in part a matter of interpretation. Is it, therefore, justifiable to call a counterfactual explanation an *explanation*? This is what we already asked. There may not be any straightforward answer, but according to Scriven [25], for instance, it might.[23] For Scriven, an explanation is something which '... fills in a particular gap in the understanding of the person or people to whom the explanation is directed'. It is a description rather than an argument, and it may even be nonlinguistic.

Whether or not a given counterfactual explanation is an explanation proper, it is evident that it fill in a particular gap in one's understanding of the scientific change involving counterfactual assumptions. This increase is many-faceted. Some aspects of it have been discussed above; let us now consider another aspect: its holistic nature. The counterfactual model of explanation represents in an exact fashion the complexity which is often characteristic of the conceptual relation of laws. Take, for example, the much discussed relation of Newton's second law and Minkowski's force law. The mere facts that the former gives good approximations of the latter for particles whose velocities are sufficiently low and that the former is the limiting case of the latter as the 'velocity of light' approaches infinity (and certain other conditions are satisfied) do not increase our knowledge concerning the complex relationship between the meanings of the two laws, or of the terms occurring in them, since those facts are purely computational. The counterfactual model, on the other hand, explicitly displays how logical features must be complemented by pragmatic ones.[24]

The positive role that translation plays in applications exemplifies this holistic nature of intertheoretic relations. That translation plays such a positive role is somewhat contrary to what Kuhn [7] maintains. Though the present notion of translation is different from Kuhn's notion—the criteria of adequacy are here less demanding, as we saw—certain detailed case studies seem to call in question the va-

[22] I shall not consider here other suggestions, but see, e.g., Bonevac [2] and Rott [23].

[23] See also Sintonen [26].

[24] For a more general discussion of such a collaboration of logical and pragmatic (and even hermeneutic) features, see my [21].

lidity of Kuhn's argument concerning the relation of translation and learning.[25] Kuhn correctly argues that, for instance, in learning Newtonian concepts of mass and force they must be learned together and that Newton's second law plays a role in their acquisition. Without recourse to the second law, one cannot learn to understand Newtonian force and mass. But then Kuhn draws the conclusion that this holistic nature of learning makes it impossible to translate Newtonian 'force' and 'mass' into the language of another physical theory in which Newton's second law is not valid, as, for instance, into the language of relativistic particle mechanics. Since they must be learned together, they cannot be translated individually.

A case study, referred to in the above, makes it clear, however, that the holistic nature of how we understand mass and force—whether Newtonian or relativistic—is a presupposition for a translation, rather than an obstacle. The translation of the Newtonian language into the relativistic one is *guided* by a holistic understanding of the relation of Newton's and Minkowski's laws. From the condition (1), in Section 2, above, we may see that a translation in our sense presupposes that certain syntactic and semantic requirements fit in with each other. What cannot be seen from (1), however, is that those requirements are determined by the force laws in the first place, in addition to the methodological and logical framework employed.

Why Kuhn thinks that the holistic nature of learning makes a translation impracticable can, perhaps, be understood in the light of his strict notion of translation. For, as we saw above, Kuhn requires that a translation must preserve reference, intension, and, perhaps, the structure of the world. On the other hand, as noticed by Pearce [11], a translation that would not bring out the conceptual disparities of the central terms of the two theories—classical and relativistic mechanics— would reproduce the syntactical inconsistency between the theories. Therefore, it seems evident that Kuhn's notion of translation is fruitless, in the sense in which the present one is not, since it cannot be positively applied so as to increase our understanding of intertheoretic relations.

The model of counterfactual explanation as well as the other models of intertheoretic explanation and reduction discussed in the literature are usually applied so as to provide a way of understanding earlier or less advanced theories or laws in the light of later or more advanced ones. Thus, for example, it is asked whether relativistic particle mechanics explains classical particle mechanics in some sense or increases or changes our understanding of the latter.

On the other hand, looking at the new theory from the standpoint of the old paradigm may result in a completely different picture of the relationship between the two theories. Historically, this is even the more natural direction of understanding in such cases of scientific change where the old theory was somehow a starting point when the new one was invented or if it is being taught first in schools and universities. Consider, for instance, the case of quantum mechanics. The theoretical, mathematical part of this theory was being developed by using some principles and concepts of classical physics as a starting point, whereby classical terms were

[25] For such case studies, see [13, 14, 23].

carried over to quantum mechanics.[26] It is evident, on the other hand, that, Bohr's complementary principle was motivated by the view that quantum-theoretical measurements must be described by means of classical terms despite the fact that quantum phenomena cannot be explained by means of classical physics. Bohr defends this view by arguing that by the word 'experiment' scientists refer to a situation about which they can tell other people what they have done and learned, and therefore experiments and their results must be reported by means of an unambiguous language in which classical terminology is being used.[27] On the other hand, Bohr argues that classical concepts can only be used in a restricted way. One cannot obtain a single complete picture of the phenomenon in question but must be satisfied with two complementary pictures. It follows, according to Bohr, that in quantum mechanics there obtains a tension between the classical and nonclassical terminology.[28]

Similar problems may occur in other fields of inquiry as well. Imagine, for example, a radical, paradigmatic change in which some crucial principles of classical logic and mathematics are replaced by other principles, say intuitionistic ones. Then we have to ask to what extent we are capable of learning to carry out our logical and mathematical thinking in the new way and how long it would take to learn it, or whether a complete change in our thinking is at all possible. Since classical terms are part of the intuitionistic language, there obtains a tension and if the change in question is universal—as it in this case would be—so as to extend to the metatheoretical level, how could one learn to live with the tension?

Since a change in our basic mathematics and logic would also comprise the metatheory, it seems that the problems involved here would in some sense be even greater than in the case of quantum mechanics, or in any changes of physical paradigms.[29] But in physics and other empirical sciences it may in some cases be equally difficult to fully adopt a new way of structuring the world because a classical term suggests a classical visualisation. The great number of different quantum-theoretical interpretations is evidently one consequence of this difficulty. When old terms, such as 'particle', 'wave', or 'mass' are employed as scientific knowledge is improved or extended to a new area, these terms are then used metaphorically, and they must be understood via how they were understood classically. Thus there is some cognitive tension as always in cases of metaphor, but to what extent this tension is cognitively fertile in science would deserve a closer study.

University of Tampere, Finland

References

[1] W. Balzer, C.-U. Moulines and J. D. Sneed. *An Architectonic for Science*, D.

---

[26] See, e.g., van der Waerden [28].
[27] See also Scheibe [24], which contains an excellent discussion of Bohr's views.
[28] See Scheibe [24].
[29] I have discussed these problems in my [17] and [19].

Reidel, Dordrecht, 1987.

[2] D. A. Bonevac. *Reduction in the Abstract Sciences*, Hackett Publishing Co., Indianapolis–Cambridge, 1982.

[3] C. Glymour. On some patterns of reduction. *Philosophy of Science*, **37**, 340–353, 1970.

[4] N. Goodman. *Fact, Fiction and Forecast*, Harvester Press, Sussex, 1979.

[5] J. Hintikka. *The Semantics of Questions and the Questions of Semantics*. Acta Philosophica Fennica, Vol. 28, North-Holland Publishing Co., Amsterdam, 1976.

[6] T. S. Kuhn. *The Structure of Scientific Revolutions*, 2nd Edition, enlarged. University of Chicago Press, Chicago, 1970.

[7] T. S. Kuhn. Commensurability, comparability, communicability. In *PSA 82*, P. Asquith and T. Nickles, eds. Philosophy of Science Association, East Lansing, 1983.

[8] T. S. Kuhn. Afterwords. In *World Changes. Thomas Kuhn and the Nature of Science*, ed P. Horwich, pp. 311–341. The MIT Press, Cambridge, MA, 1993.

[9] D. Lewis. *Counterfactuals*, Blackwell, Oxford, 1973.

[10] D. Pearce. *Translation, Reduction and Equivalence: Some Topics in Intertheory Relations*, University of Sussex, 1979.

[11] D. Pearce. *Roads to Commensurability*, D. Reidel, Dordrecht, 1987.

[12] D. Pearce and V. Rantala. New foundations for metascience. *Synthese*, **56**, 1–26, 1983.

[13] D. Pearce and V. Rantala. A logical study of the correspondence relation. *Journal of Philosophical Logic*, **13**, 47–84, 1984.

[14] D. Pearce and V. Rantala. Limiting-case correspondence between physical theories. In *Reduction in Science*, W. Balzer, D. Perace and H.-J. Schmidt, eds, pp. 153–185. D. Reidel, Dordrecht, 1984.

[15] D. Pearce and V. Rantala. Approximate explanation is deductive-nomological. *Philosophy of Science*, **52**, 126–140, 1985.

[16] V. Rantala. The old and the new logic of metascience. *Synthese*, **39**, 233–247, 1978.

[17] V. Rantala. Scientific change and change of logic. In *Intensional Logic, History of Philosophy, and Methodology. To Imre Ruzsa on the Occasion of his 65th Birthday*, I. M. Bodnár, A. Máte and L. Pólos, eds., pp. 247–252, Budapest, 1988.

[18] V. Rantala. Counterfactual reduction. In *Imre Lakatos and Theories of Scientific Change*, K. Gavroglu, Y. Goudaroulis and P. Nicolacopoulos, eds. Kluwer Academic Publishers, Dordrecht, 1989.

[19] V. Rantala. Reduction and explanation: Science vs. Mathematics. In *The Space of Mathematics. Philosophical, Epistemological, and Historical Explorations*, J. Echeverria, A. Ibarra and T. Morman, eds., pp. 47–59, Walter de Gruyter, Berlin, 1992.

[20] V. Rantala. Translation and scientific change. In *Theories and Models in Scientific Processes*, (Poznan Studies in the Philosophy of the Sciences and Humanities, Vol. 44), W. E. Herfel, W. Krajewski, I. Niiniluoto and R. Wójcicki, eds., pp. 249–268, Rodopi, Amsterdam, 1995.

[21] V. Rantala. Explanatory translation and conceptual change. In *Methods of Reading*, I. Koskinen, E. Oesch and T. Vadén, eds. Tampere, 1995.

[22] V. Rantala and T. Vadén. Idealization in cognitive science. A study in counterfactual correspondence. In *Idealization VII: Structuralism, Idealization and Approximation*, (Poznan Studies in the Philosophy of the Sciences and Humanities, Vol. 42), M. Kuokkanen, ed., pp. 179–200, Rodopi, Amsterdam, 1994.

[23] H. Rott. On relations between successive theories. In *Abstracts of the LMPS 87*, Vol. 4, Part 2, pp. 123–127. Moscow 1987.

[24] E. Scheibe. *The Logical Analysis of quantum Mechanics*, Pergamon Press, Oxford, 1973.

[25] M. Scriven. Explanations, predictions and laws. In *Minnesota Studies in the Philosophy of Science*, Vol 3, pp. 170– 230, H. Feigl and G. Maxwell, eds. University of Minnesota Press, Minneapolis, 1962.

[26] M. Sintonen. *The Pragmatics of Scientific Explanation*. Acta Philosophica Fennica, Vol. 37, Helsinki, 1984.

[27] R. Tuomela. *Theoretical Concepts*. Springer-Verlag, Wien-New York, 1973.

[28] B.L. van der Waerden. Introduction. In *Sources of Quantum Mechanics*, B. L. van der Waerden, ed., Dover Publications Inc., New York, 1967.

2

Genetically-Constructive Ways of Theory Building

Vyacheslav Stiopin

Logico-methodological literature traditionally regards a natural-science theory as a hypothetico-deductive system. This view manifests itself in two different forms. According to what is known as the strong version which has developed within the framework of the so-called standard concept, a natural-science theory can be considered as structurally similar to interpreted calculi or informal axiomatic theories in mathematics. The strong version views the unfolding of a theory in the course of an explanation and forecasting of new facts as a process of logically deducing theoretical corollaries from axioms, with lower-level propositions being inferred in a strictly logical manner from basic upper-level statements of the theory until propositions are formulated which can be compared with experimental data.[1]

Critics of the standard concept have developed what is referred to as the weaker version of the hypothetico-deductive model. This new version takes into account the possibility of expanding the scope of, and defining more clearly, the initial concepts of a theory as the latter unfolds itself. In this case of its interaction both with experience and other theories. It is viewed as a deduction of conclusions from basic laws and principles, which allows of new assumptions and makes it possible to expand and specify the original theoretical suppositions.

The weaker version is more in keeping with actual research practices than the strong one, yet this general definition gives no idea of the mechanisms whereby a theory unfolds and evolves.

Rather than merely stating the fact that a functioning theory allows of new assumptions, an important thing in this connection is to find out how and when assumptions are made, whether there are any, albeit undisclosed, norms to regulate that process, and if such norms really do exist, what they boil down to. This problem can be solved with the aid of a special analytical procedure giving a clearer idea of general concepts of the weaker version of the hypothetico-deductive model. first and foremost, it is necessary to obtain a clearer view of the unfolding of a theory in the course of its functioning, i.e. in the course of an explanation and forecasting of

[1] Braithwaite, [1].

P. I. Bystrov and V. N. Sadovsky (eds.), Philosophical Logic and Logical Philosophy, 17–31.

new facts. To achieve that goal, one should begin analysing some actual natural-science theories.

1 The unfolding of a theory as a solution of theoretical problems

In view of the fact that theoretical knowledge falls under the category of complex developing systems, it is best to examine the laws of the functioning and development of theoretical knowledge at higher levels of its historical evolution. Considering that physics is the first empirical science to have brought forward a broad range of nature theories, one would be well advised to refer to physics in order to observe the specific features of the functioning of a natural-science theory. A close look at physical theories makes it clear that the unfolding of a theory does not confine itself to a strictly logical deduction of one proposition from another. Although the procedures of a logical deduction of theoretical propositions and of a displacement in mathematical formalism are an important part of the process of making theoretical inferences, they do not constitute the whole of that process. A special role in the development of theories related to empirical sciences is played by genetico-constructive operations in which reasoning takes the form of mental experiments performed on abstract objects of a theory.

In Soviet Science literature, the difference between genetico-constructive operations and the development of axiomatic theories was first noted by V. A. Smirnov in the early 1960s.

Contrary to the axiomatic method, when 'a system of propositions, describing a certain realm of objects and a system of logical operations over propositions are taken as a primitive basis', the genetic method implies operations directly with abstract objects of the theory which are fixed in corresponding signs. The process of reasoning in this case takes the form of mental experimentation over abstract objects 'taken as concretely present'.[2]

One of the examples of such development of the theory is the Euclidean geometry. V. A. Smirnov demonstrated that the traditional interpretation of Euclid's 'Foundations' only as an example of axiomatic construction of the theory does not take into account the genetic-constructive methods which are used there. "The conception of 'Foundations' ", writes Smirnov, "is not an excellent attempt to reach the ideal of the axiomatic method in its modern understanding at constructive (genetic) building of a theory".[3] Euclidean postulates introduced major abstract objects—point, line, circle, segment— as being defined through construction with the help of an ideal compass and ruler, various geometrical figures out of major objects. Mental experiments with figures (their decomposition and transformation) served as a basis for attaining knowledge fixed in the system of corresponding propositions of

[2] Smirnov [4, p. 269].
[3] Smirnov [4, p. 278].

the Euclidean geometry.

In the late 1960s to early 1970s I analysed the structure and functioning of the physical theory from the point of view of genetically-constructive method.

The analysis revealed that the functioning of a theory does not only mean that some abstract notions give rise to others, but is also due to the presence of special organization levels in the system of theoretical objects and to a transition from one level of theoretical objects to another. An analysis of the internal structure of each of these levels and of the relationship between them made it possible to distinguish special subsystems of theoretical constructs which serve as a basis for the development of a theory. Such subsystems of constructs represent the essential characteristics of the reality reflected by a theory and are models of objects and processes with which the theory is concerned.

The hierarchy of theoretical models constitutes the internal skeleton of those lower levels. The reduction itself is the emergence of models on the basis of other models in the process of mental experimentation with theoretical constructs and the adoption of limitations and particularizations in regard to a given model.

Consider a simple example which illustrates the principle features of the above process. Let it be assumed that one has to derive the law of small-amplitude oscillation from the equation describing Newton's second law. A specialist in mechanics can do it in different ways. The first can be traced back to Euler's works on mechanics and is frequently mentioned in present-day textbooks. This method calls for particularizing the type of force in the equation which describes Newton's second law. To make that possible, one has first to explicitly formulate the physical meaning of the equation. As with others of Newton's laws, the second law and its equation refer to an idealised model which reflects the essence of any mechanical process. This model refers to a displacement of a material point within a reference system and to changes in its motion resulting the effects of a force. It is from this standpoint that a researcher examines mechanical oscillation as a specific type of mechanical process. The researcher regards the motion of an oscillating body as a displacement of a material point within a reference system. Having noted the specifics of empirically observed events of oscillatory motion, he performs a number of mental experiments with objects of the basic theoretical model. Thus he finds out that on being set in oscillatory motion, a material point periodically returns to a state of equilibrium, whereby the type of force acting on that point is individualized and the abstract object referred to as 'force' is replaced by a new object referred to as 'restoring force'. The researcher can use the basic theoretical model to design an oscillator as a specific model of mechanical oscillation. The equation of motion of a material point, i.e. Newton's second law, is then applied to the oscillator, whereby another equation is produced which describes the law of oscillatory motion (in the equation $F = m\ddot{r}, kx$, which is the restoring force, is substituted for F, the resultant equation being $m\ddot{x} + kx = 0$).

The foregoing procedure of drawing inferences from basic laws of a theory is universal, considering that the unfolding of even the most developed and highly mathematized theories of physics necessarily calls for mental experimentation with theoretical models.

Such models are not external in relation to a theory, because they are part of the theory and make up its internal skeleton. They are to be distinguished from analogue models which serve as a scaffolding, as it were, for constructing a theory, but are now wholly incorporated in that theory. It is suggested by this author that theoretical models incorporated in a theory be called theoretical diagrams, for these are indeed diagrams of a given theory and reflect the essential relations of those objects and processes.

Within a theory, one has to distinguish between a basic theoretical diagram and its specific versions. From the viewpoint of the internal structure, these are small sets of theoretical constructs (abstract objects) related to one another in a strictly definite manner. Coming back to Newton's mechanics, we have three basic abstract objects which make up its basic theoretical diagram. These are: the material point, the force and the reference system. From the standpoint of construction, each of the three basic abstract objects is independent, i.e. none of these can serve as a basis for constructing another within the formulatory framework of the theory. At the same time, a basic diagram can be used for constructing specific theoretical diagrams.

The relationships between basic abstract objects are described by the laws of the theory they are related to. The relationships between abstract objects of specific theoretical diagrams are expressed by specific theoretical laws, such as the laws of mechanical oscillation, rotation of bodies, motion of a body in a central force field, etc. Equations in which laws of physics find a mathematical expression can be interpreted due to their relationship with theoretical diagrams. The values of individual terms of an equation directly express features of constructs of related theoretical diagrams, whereas a solution of an equation can be regarded as a special way of operating with given constructs.

In mathematical formalism, combinations of original values incorporated in an equation produce new values. From the standpoint of operating with abstract objects of a theoretical diagram, this means that new features are derived from basic features of objects and can be abstracted as special constructs. For example, the solution of equations describing Newton's laws discloses such properties of the material point as its kinetic and potential energy, and momentum conservation, etc. The new properties, i.e. momentum, kinetic and potential energy and so on can be used as relatively independent abstractions representing essential characteristics of mechanical motion. Outside mechanics, these abstractions can be used in describing other physical processes.

Thus the solution of informal physical equations can be regarded as an investigation of a theoretical diagram and a disclosure of implicit information on reality contained therein. This information is rendered explicit by generating new abstractions on the basis of the original abstract objects of the theoretical diagram.

In operating with equations, new abstractions are formed with respect to both the basic and specific theoretical diagrams. Thus at each level of the hierarchy of abstract objects of a theory, we see both relationships of principal objects, which make up theoretical diagrams, and a network of 'sister' abstractions associated with each theoretical diagram. Each such network is a special subsystem within a developing system of abstract objects related to a given scientific theory. These subsystems of abstractions combine into a single whole, because they all hang on the framework of the theory, which is the hierarchy of theoretical diagrams with new specific theoretical diagrams continuing to develop from the basic one. Considering that a transition from fundamental equations of a theory to mathematical formulations of specific theoretical laws is invariably accompanied by corresponding transformations of theoretical diagrams, it can be stated that the interpretation of the body of mathematics related to a given theory involves the entire set of theoretical diagrams and their associations with experience.

The development of this body of mathematics can be likened to the development of a calculus only to a certain degree, because only some individual fragments of that body stem from the derivation of some formulas from others in accordance with the rules of mathematics. These fragments are brought and held together by conducting mental experiments with theoretical diagrams. When necessary, these are rendered explicit and presented in the form of special model concepts which are recorded either as drawings complete with explanatory notes or as descriptions of properties and relationships of constructs incorporated in the theoretical diagram. It is precisely through mental experimentation with such constructs that the basic equations of a theory are particularized to suit a particular physical phenomenon and that specific theoretical laws are formulated to describe the given situations.

In the light of their relatively independent status, specific theoretical diagrams and laws related thereto can be viewed as a nucleus of a special theory. This view is justified not only by the fact that such diagrams are independent of one another, as, for example is the case with the theoretical diagrams of mechanical oscillation and rotation of bodies, although each of these can be constructed on the basis of the basic theoretical diagram of mechanics. Such specific diagrams are related to a specific area, have specific applications and can exist before they are incorporated into a developed basic theory.

This writer has analysed cases of an independent existence of specific theoretical diagrams and laws as a nucleus of a special scientific theory as well as peculiarities of their change upon being incorporated into a generalizing basic theory.

An examination from this standpoint of a mature basic theory makes it clear that the specific theoretical diagrams which develop in the course of its unfolding and the subsequent applications of these diagrams to describing and explaining concrete empirical situations can be regarded as the generation by the basic theory of special theories that can also be called microtheories. It is important that researchers be able to distinguish between two types of such theories which differ in the nature of the theoretical diagrams upon which they are based. Special theories of the first type may be fully incorporated in a generalizing basic theory and constitute a whole part thereof. The laws and models of small oscillation, of rotation of solid bodies, and so on, which are incorporated in mechanics, are only partially related to a single basic theory. The theoretical diagrams which form their basis are a kind of hybrid formation, as stem from basic theoretical diagrams of at least two theories.

An example of such a hybrid formation is the classical model of black-body radiation which has developed from concepts of thermodynamics and electrodynamics. Hybrid theoretical diagrams may exist as independent theoretical constructions on a par with basic theories and non-hybrid specific diagrams not yet included in the basic theory.

This complex system of interacting of basic and specific theories makes up the body of theoretical knowledge of a given science.

Each theory, even a special one, is structured as a hierarchy of theoretical diagrams. Bearing that in mind, the division of theoretical diagrams into basic and specific is relative and only makes sense in dealing with a given theory. For example, as a model of mechanical oscillation, the harmonic oscillator is a specific diagram relative to the basic theoretical diagram of mechanics. At the same time it has the status of a theoretical diagram in relation to more specific theoretical models which serve to describe specific types of mechanical oscillation, such as degenerate oscillation of a pendulum, damped oscillation of a pendulum or a spring-loaded body, and so on.

In drawing inferences from basic equations of any theory, be it a basic or a special theory (microtheory), the researcher mentally experiments with theoretical diagrams, using particularizing assumptions and reducing the basic diagrams of a given theory to an appropriate specific theoretical diagram.

Complex forms of theoretical knowledge, such as the theory of physics, are distinct from other forms of knowledge in that the operations of developing specific theoretical diagrams on the basis of constructs of the basic theoretical diagram cannot be clearly described in postulates and definitions of the theory. Such operations are illustrated by what can be termed as reference situations which are incorporated in the theory and show how inferences are drawn from the basic equations of the theory. In mechanics, examples of such reference situations are found in the derivation from Newton's laws of the laws of small oscillation, motion of bodies in the central force field, rotation of solid bodies, and so on. In the classical electromagnetic field theory, a case in point is the derivation of the laws of electromagnetic and electrostatic induction and some other laws from Maxwell's equations and the Biot–Savart, Coulomb and Ampere laws. The informal nature of these procedures and the need for taking into account the specifics of the object under investigation while constructing specific theoretical diagrams turns the deduction of each corollary from basic equations into a special theoretical problem. Thus each theory unfolds as such problems are solved. The solution of some of them serves as an exemplar for solving other related problems.

A natural question arises in this connection; where do original exemplars come from and how are they incorporated into a theory?

A special role of exemplars in the functioning of a theory was noted by Thomas S. Kuhn who insisted that these were the most important part of the paradigm of 'ordinary research' and that their structure and applications needed a careful analysis. He pointed out that the use of exemplars in ordinary research is similar to the development of exemplars in the history of science. The similarity lies in drawing analogies between different and often seemingly incompatible physical situations which are examined from a common standpoint.[4]

In the early 1970s, Kuhn's ideas in regard to the functioning of models in the process of problem solving were evolved to a certain extent by I. D. Sneed, W. Stegmüller and other proponents of what is known as the structuralist approach to the analysis of a scientific theory.[5] These researchers noted some characteristics of the structure and functioning of the physical theory which were discovered in the same period by their counterparts in the USSR and descried in Soviet methodological literature. It was found that the mathematical body of a theory is built around a plurality of coordinated models. It was also established that the unfolding of a theory is introducing limitations into models and thus actualizing a certain subset of the entire set of possible models of the theory applicable to actual physical situations.

Finally, an important part of the Sneed–Stegmüller concept is the idea of the unfolding of a basic theory as a generation of microtheories whereof each is a certain theoretical element generated by a basic theoretical element (the latter being a combination of fundamental laws related to the basic model, from which specific laws are derived by introducing certain limitations into the basic element). Thus exemplars of problem solutions are linked with the problem of reducing basic theoretical elements to lower-level elements. However, the structuralist concept fails to specify the operations which characterize the introduction of limitations whereby the aforesaid reduction is made possible. It also fails to cast light on the origin of exemplars of problem solutions.

In this author's opinion, it is not accidental that the Sneed–Stegmüller concept gives no answer to these crucial questions. The views of these scientists in regard to the structure of a theory as an integral system of theoretical elements are largely applicable to a global phenomenological description of the integrity of a theory. On the other hand, they do not pay enough attention to the details of the theoretical organization, which bear on the analysis of the microstructure of theoretical models, as well as their relations with one another and with the reality they reflect. Yet it is precisely a detailed study of the internal structure of theoretical models and of procedures of vindicating them as schematic representations of actual physical interactions that is needed for a clear understanding of the structure and formation of problem solution exemplars.

Proceeding from the above concepts of the role of theoretical diagrams in the functioning of a theory and in the solution of theoretical problems, I shall try to answer the question of how primary problem solution exemplars develop within a theory. The problem of their origin can be formulated as that of the genesis of theoretical diagrams.

---

[4] Stiopin, [7, 8, 9, 10].
[5] Stiopin, [11].

2 The establishment of problem solution exemplars in the genesis of a theory

The process of the formation of theoretical diagrams can best be examined through an analysis of concrete historical material. With that object in view, L. Tomilchik and this writer have traced back the history of Maxwell's electrodynamics. We chose that subject for two reasons.

First, Maxwell's theory plays a special role in the history of physics, as it belongs to, and completes the classical period in the development of that science. At the same time, that fundamental theory already manifests some theory-building methods typical of present-day physics. This makes it possible to identify theory-building operations common for both the classical and present-day periods in the development of physics with due regard for the specifics of each of these periods.

Secondly, we were attracted by the unique nature of the historical documentation related to Maxwell's discovery. It presents an absolutely clear picture of the formation of a scientific theory, for all the research that had led to that theory was conducted by one man whose writings record all the intermediate versions of his theory, including those that were later rejected by Maxwell himself.

The emergence of Maxwell's theory was preceded by an acquisition of a body of theoretical knowledge which reflected the essential characteristics of only some individual aspects of electrogenetic interaction. That body of knowledge included the laws discovered, and theoretical models developed by Coulomb, Faraday, Ampère, and other scientists. Those were specific theoretical laws and specific theoretical diagrams in relation to the electromagnetic field theory developed by Maxwell. In fact, the latter theory came as the result of a consistent generalization and synthesis of the said specific theoretical laws and diagrams. Such a way of developing a fundamental physical theory is a rule rather than an exception from the rule, considering that all the fundamental theories of classical physics, including Newton's mechanics, thermodynamics, the classical electromagnetic field theory, resulted from specific theoretical laws and models which were later generalised by respective fully-fledged theories. A theoretical synthesis of a similar type can also be found in the history of quantum physics.

Maxwell's synthesis was based on the use of well known analogue models borrowed from continuum mechanics and serving as equations which he incorporated in his electromagnetic field theory.

The use of analogies deserved a careful analysis, for it is a universal way of developing a new theory. First and foremost, it will be noted that this operation involves the use of mathematical structures and concepts of mature theories as means for developing a new theory. Physical theories are never isolated, but developed as a system in which some theories provide building blocks, as it were, for the development of others. What is also important, the analogue models used by Maxwell, such as incompressible fluid stream tubes and compressible vortices, are theoretical diagrams of continuum mechanics.

When equations related to these diagrams were translated into the language of electrodynamics, old mechanical values were substituted by new values. The abstract objects of mechanics incorporated in the analogue model were replaced by new objects, such as lines of force, charges, infinitesimal current elements and so on. These latter objects were borrowed by Maxwell from the theoretical diagrams of Coulomb, Faraday and Ampère, which he generalised and incorporated into his new theory. A researcher is not always aware of a substitution of new objects for old ones in an analogue model, yet such a substitution necessarily takes place in developing a new theory, otherwise old equations are devoid of a new physical content and cannot be applied to new areas of physics. The substitution means that abstract objects borrowed from one area of knowledge (in our example this area is the system of information on electricity and magnetism) are combined with a new structure, referred to as a 'network of relationships', which is borrowed from another field of knowledge (which in our example is continuum mechanics).

The merger leads to a transformation of the analogue model which turns into a theoretical diagram related to a new field of knowledge. In its original form, this diagram is merely hypothetical and needs corroboration and vindication. The latter process is the most important thing in the development of a theory, because upon combining with a new network of relationships, abstract objects normally acquire new attributes. If the researcher feels that an analogue model thus constructed represents essential features of a new object of research, he makes two assumptions. First, he assumes that the new hypothetical features of abstract objects are related precisely to those empirically observable phenomena which his model is called upon to explain. Second, he assumes that these new features are compatible with other essential features of abstract objects, which are known from previous research or practice. Of course, the truth of such assumptions must be proven, which is done by using abstract objects as idealizations based on the latest experience. The attributes of abstract objects, which are hypothetically introduced 'from above' in relation to experiments in the new area of physical interaction, are now reproduced 'from below'. This is done by conducting mental experiments corresponding to typical characteristics of actual experimental situations which the theoretical model is expected to explain. The final task is to make sure if the new attributes of the abstract objects agree with those known from previous experience.

The purpose of this set of operations is to corroborate the attributes of the abstract objects incorporated into the hypothetical model and transform the latter into a theoretical diagram related to a new area of physical interaction. Such operations can be referred to as a constructive introduction of objects into a theory. It is also suggested that a theoretical diagram which is in agreement with the above procedures be referred to as constructively vindicated.

Constructive vindication relates theoretical diagrams to experience, which means that it relates to experience physical quantities of the body of mathematics of a given theory. It is precisely due to constructive vindication procedures that the rules of correspondence are developed within a theory.

In pre-Maxwellian physics, a constructive vindication of theoretical diagrams was no problem. Being largely implicit, it was often effected without the researcher being aware of it. This is not so in Maxwell's work, especially at the final stage of his work on a single theory of electromagnetism. In modern physics, constructive vindication procedures manifest themselves quite clearly and are even regarded as a special area concerned with empirical interpretation of mathematical formalisms. The well known Bohr–Rosenfeld procedure in quantum electrodynamics are a case in point.

The role of constructive vindication in the development of a theoretical diagram and in an adequate interpretation of equations can be illustrated by a description of an important phase of Maxwell's work. Consider his first steps in the development of the electromagnetic field theory, the period when he produced a generalizing formulation of the laws of electrostatics. He discovered those laws by using the hydrodynamic analogy of incompressible fluid stream tubes. Substituting electric lines of force for the stream tubes in the analogue model, he constructed a generalised hypothetical diagram of electrostatics and presented Euler's equations as a description of the behaviour of the lines of force. As a result of the introduction into the analogue model of abstract objects borrowed from Faraday's model of electrostatic induction, these objects, i.e. the lines of force, were placed within a new network of relationships and thus acquired new attributes. The lines of force were 'isolated' from the changes which produced them. Potentially, this was a new, albeit hypothetical, idea of an electric field independent of its sources.

The hypothesis of an independent existence of electric lines of force could be raised to the status of a theoretical statement only on condition that an empirical proof of the new attribute of the lines of force was found. To find that proof was not a difficult job, considering the possibility of another mental experiment with Faraday's concept of electrostatic induction. The next logical step was a consideration of the ideal capacitor in which the lines of force are viewed as those produced in an idealised dielectric material placed between ideal charged plates. In this case the lines of force are dependent on the magnitude of the plate charges. By mentally varying the plate charges and observing the resultant increases or decreases of the amount of electrical energy in the dielectric material, the limiting case was eventually encountered when all the electric energy was concentrated in the dielectric material. That case corresponded to the concept of a set of lines of force which continue to exist even after the removal of the charges which produce them. 'Isolated' from the charges, such lines of force become an idealization based on experimental findings.

As a result, Maxwell was able to continue with his theoretical synthesis and present Coulomb's law in field terms. Thus a charge could be determined on the basis of the characteristics of those generated by that charge.

It is clear from the above that the development of theoretical systems for the interpretation of physical equations does not only mean the establishment of analogies between different physical situations, but also includes the constructive vindication procedure.

The primary patterns of tasks decisions arise in the theory genesis just in the interaction of these two operations.

The procedures of constructive substantiation are not in the center of Kuhnian concept or of Sneed's and Stegmüller's works. It is not wonderful that Kuhn and his followers are accentuating only one side of the question—the role of analogies as the base tasks decision. The operations of substantiation arising in this process of theoretical schemes fall out of the sphere of their analysis.

It is quite significant that in the limits of this approach the principle difficulties in the attempts at clarifying which is the role of correspondence rules and what is their origin arise. Kuhn, for example, suggests that in the activity of the scientific community these rules do not play such an important role, which the methodologists ascribe to them traditionally. He underlines that the search for analogies between different physical situations and the use of the already found formulae is the most important aim in the tasks decision. As to the correspondence rules Kuhn considers them the result of the following methodological retrospection when a methodologist tries to ascertain the criteria which the scientific community is using by applying one or another analogy.[6] Generally speaking, Kuhn is consistent in his position because the question on the procedures of constructive substantiation for theoretical patterns does not arise in the limits of his conception. For displaying this procedure a special approach to the structure and the dynamics of scientific knowledge is demanded: it is expedient to consider the theoretical models which are included into the theory simultaneously as an ontological scheme which reflects the essence characteristics of the examined reality and as an original 'rolling up' of the matter-practical procedures in the limits of which the mentioned procedures may be revealed principally. Just this vision allows a finding and describes the operations of theoretical schemes of constructive substantiation.

But as soon as the constructive substantiation ensures the appearance of the correspondence rules in the theory defining their content and meaning Kuhn's difficulties in the definition of the ways of forming and functioning of these rules become common.

It is typical that Kuhn makes reference to Maxwell's history of electrodynamics discussing the problem of patterns. Analysing it only in the sphere of the analogy models usage he supposes that the main results of Maxwellian research were obtained without any construction of the correspondence rules.[7] But we may scarcely accept this conviction. The matter is that at one of the stages during the building process of his theory Maxwell had obtained the field equations very close to the modern mathematical scheme of electromagnetic phenomena descriptions. However, he was not capable at that period of time to put into accordance the real attitudes of the objects of empirical situations with some fundamental magnitudes (the theoretical scheme introduced together with the equations did not find the constructive substantiation). And then Maxwell had to abandon this general perspective apparatus and began anew the process of theoretical synthesis. In this pro-

[6] Kuhn, [2].

[7] Sneed, [5], Stegmüller [6].

cess the search for mathematical structures describing electromagnetic interactions was constantly supported by explication and the basing of introduced theoretical schemes.

Tracing back the history of the development of Maxwellian electrodynamics with due regard for the foregoing operations incorporated in the construction of theoretical models leads one to an understanding of the logic of Maxwell's theoretical synthesis. What the great scientist did was a step-by-step generalization of his predecessors' theoretical knowledge of individual types of electromagnetic interaction. The theoretical material which he generalized was naturally divided into different areas of knowledge, such as electrostatics, magnetostatics, steady-state current, electromagnetic induction, and force and magnetic effects of steady-state currents.

By using analogue models, Maxwell first developed generalized equations for individual bodies of knowledge. In the course of that process, he also constructed a generalizing hypothetical model for the interpretation of the equations and the assimilation of theoretical diagrams related to a given body of knowledge.

Following a constructive vindication of that model and having transformed it into a theoretical diagram, Maxwell turned to another area of knowledge in his generalization. He used a previous hydrodynamics of mechanics analogy which he updated and made more sophisticated to assimilate new physical material. The events of the vindication procedure described above were then repeated: a new constructive content was equivalent to the explication of a new generalizing theoretical diagram. It was proven that the new diagram could serve to assimilate specific theoretical models of the new body of knowledge, and that specific theoretical laws could be deduced from the new generalizing equation. However, the vindication procedure did not end at that point.

The scientist had to make sure that his new generalization did not erase the previous constructive content. For that purpose he again deduced all the specific laws of the previously synthesized bodies of knowledge from his latest generalizing equations.

It is significant that in the course of the latter process each new generalizing theoretical diagram was reduced to specific theoretical diagrams equivalent to the previously assimilated ones.

At the last stage of his theoretical synthesis, when the principle equations of the theory had been developed and the basic theoretical model had been completed, Maxwell presented the final proof of the correctness of the equations he had introduced, as well as of their interpretation. He used the basic theoretical model to construct respective specific theoretical models and deduced, in a new form, all the specific theoretical laws from the basic equations in which those laws were generalized. It was proven at the final stage of the development of Maxwell's electromagnetic filed theory that the theoretical model of the electromagnetic field could lead to such specifics as the theoretical diagrams of electrostatics, direct current, electromagnetic induction, etc. It was also proven that from the electromagnetic field equations, one could deduce Coulomb's, Ampère's and Biot-Savart's laws as well as Faraday's laws of electrostatic and electromagnetic induction.

The last stage of Maxwell's synthesis can also be regarded as an exposition of a 'finalized' theory. The process of the development of that theory is now reversed by unfolding the theory and deducting theoretical corollaries from the principal equations. Each of these deductions can be viewed as a method for, and the result of, solving theoretical problems. While being an indispensable element of the vindication of the theory, the informal operations now serve a new function: they become reference operations which help the researcher solve new theoretical problems. Thus, reference solutions of problems become part and parcel of the genesis of a theory.

The described way of pattern formation is invariant both in connection with classical or modern physics. But at each such stages it is being realized in a specific form.

Modern ways of physical theory building have some peculiarities which we need to take into account in discussing the patterns problem. The first of them is that a wide use of mathematical hypothesis method allows for a quite some time to go by in the search for some more new generalizing equations without applying at every step intermediate interpretations connected with the theoretical schemes constructive substantiation. This of course, does not mean that the intermediate interpretations providing for the control over the physical sense of the introduced equations stop playing an essential role in the theory building. They are as necessary as before but their number is reduced because the search for generalizing equations is being fulfilled by more large steps than in a classical period.

The hypothetic variants of the theoretical scheme implicitly introduced in the process of a mathematical hypothesis at the stage of interpretation get their constructive substantiation.

This procedure completes the forming of the patterns of theoretical tasks decisions but it also has its peculiarities in modern physics. Earlier I analyzed them on the concrete historical material of the history of quantum electrodynamics. The analysis demonstrated that in the process of empirical interpretation of equations which were received after a series of mathematical hypotheses at the level of interpretation a study goes in a condensed form through the main stages in the development of mathematical apparatus. The Bohr–Rosenfeld procedures which ensured the empirical interpretation of the apparatus of quantum electrodynamics constructed in the 1930–1940s are a bright example of this process. The logic of their enrollment coincided in its main features with the logic of the development of the theory's mathematical formalism: first the theoretical scheme of a quantum electromagnetic field was constructively substantiated; then the same was made in regard to quantification of sources (of the charge–current distribution) interacting with the radiation field, and, finally, the scheme of this interaction in the first approximation of the perturbation theory was substantiated.

The constructive substantiation of theoretical schemes in the process of a series of intermediate interpretations, as a rule, gives a new impetus to the subsequent mathematical synthesis and completing construction of the theory. In this regard the history of quantum electrodynamics may serve as quite a characteristic example, since only after N. Bohr's and I. Rosenfeld's substantiation of the theory's formalism (constructed in the 1930–40s) the so-called second birth of quantum electrodynamics connected with the working out in the 1950s of the method of restandardization and the appearance of examples of tasks which characterize the process of interaction of quantum electromagnetic and electron–positron fields in the highest approximations of the perturbation theory became possible.

Thus, while in classical physics a developed fundamental theory was constructed according to the scheme: $\text{equations}_1 \rightarrow$ intermediate interpretation_1, $\text{equation}_2 \rightarrow$ intermediate $\text{interpretation}_2 \ldots$ generalizing system of equations $\rightarrow$ generalizing interpretation; in modern physics it is a different way. First equation_1 $\rightarrow \text{equation}_2$, etc., and then—a series of intermediate interpretations, repeating in a condensed form the main stages of the apparatus' construction(interpretation_1 $\rightarrow \text{interpretation}_2$, etc.). Then may follow a new stage of the development of mathematical apparatus and a new series of interpretations which completes the formation of the theory. At the stage of interpretation preliminary variants of theoretical schemes may be fundamentally reconstructed, as some non-constructive elements are removed. In the process of this reconstruction the study repeatedly moves from theoretical schemes to experience situations and back. And this forms the methods of reduction of theoretical schemes of high levels to low levels due to making certain restrictions on them, prompted by real experience situations.

Once the theory is constructed its future life is connected with its development in the process of expansion of its field of application.

The functioning of an established theory and its growing applicability lead to the emergence of new reference solutions of problems which are included in the theory on a par with those introduced in the course of the development of the theory. The primary reference solutions also change to suit the growing scope of scientific knowledge and modifications in the previous form of the theory, yet they usually persist, in their modified form, in all subsequent expositions of the theory. Even the latest exposition of classical electrodynamics points to the application of Maxwell's equations to specific situations encountered in physics, as illustrated by the deduction of Coulomb's, Biot–Savart's and Faraday's laws from Maxwell's equations. This is one of the specifics of the functioning of a theory as a 'hereditary system'. This is not merely an apt expression, but also a succinct exposition of a concept brought forward by A. I. Rakitov. The term that a theory contains traces of its past history, and reproduces essential features of the processes of its formation as standard problems and standard methods for solving those problems.

The genesis of a theory is dialectically reflected in its organisation and determines its functioning.[8]

Russian Academy of Sciences, Moscow

---

[8]Kuhn, [3].

References

[1] R. B. Braithwaite. *Scientific Explanation: a Study of the Function of Theory, Probability and Law in Science*. Cambridge, New York, 1960.

[2] T. Kuhn. *The Structure of Scientific Revolutions*, Chicago, 1970.

[3] T. Kuhn. Second thoughts on paradigms. *The Structure of Scientific Theories*, pp. 459–482, Urbana, 1974.

[4] V. A. Smirnov. The genetic method for developing a scientific theory. *The Phisophical Problems of Formal Logic*, Moscow, 1962. (In Russian.)

[5] I. D. Sneed. Theoretization and invariance principles. In *The Logic and Epistemology of Scientific Change*, pp. 130–178, Amsterdam, 1979.

[6] W. Stegmüller. A combined approach to the dynamic of theories. *Theory and Decision*, **9**, 39–75, 1978.

[7] V. S. Stiopin. The subject and object problem in empirical science. *Voprosy filosofii*, **1**, 1970.

[8] V. S. Stiopin. Genesis of theoretical models of science *Filosofskije nauky*, **3**, 1971.

[9] V. S. Stiopin. To the problem of the scientific theory structure and genesis. *Philosphy. Methodology. Science*, **M**, 1972.

[10] V. S. Stiopin. *The Scientific Theory Formation*, Minsk, 1976. (In Russian.)

[11] V. S. Stiopin. *The Development of a Scientific Theory*, Minsk, 1976. (In Russian.)

3

Toward Holistic Conception of Artificial Intelligence

Vadim N. Sadovsky

This paper suggests a substitution to approach to the construction of a holistic conception of AI for the so far dominating narrowly special (computer, psychological, and the like) approaches to artificial intelligence (AI) problems. In the former case, the AI domain comprises at least six interrelated blocks:

1. functioning of natural intelligence;
2. development and implementation of intelligent computer programs;
3. natural intelligence theory;
4. intelligent programming theory;
5. systems-cybernetic paradigm;
6. philosophy, methodology, and logic of knowledge, primarily of scientific knowledge.

These blocks make up a wholeness, and it is only the latter that permits a productive statement of AI research problems.

This paper discusses a holistic AI concept which serves as a foundation for exploring a number of crucial AI research problems.

1 Artificial Intelligence Research Domain

The AI-related research is one of the few scientific and technological disciplines whose date of inception is known almost up to a day's accuracy. At the same time, which was often the case in the history of science, there were at least two, and not one, dates of AI origin.

Indeed, the term 'artificial intelligence' was first introduced in the summer of 1956 when John McCarthy, a prominent American computer theoretician and practitioner, hosted in Dartmouth (USA) many 'godfathers' of cybernetics—C. Shannon, M. Minsky, H. Simon, A. Newell, *et al.*—to discuss an AI project's feasibility.

P. I. Bystrov and V. N. Sadovsky (eds.), Philosophical Logic and Logical Philosophy, 33–47.

The term 'AI' was present in the very title of the Conference—Dartmouth Summer Research Project on Artificial Intelligence—and very soon it gained wide acceptance in the scientific community.

The Conference attendees could not ignore an earlier paper directly relating to AI problems (though this term was missing from the paper), notably an article of a prominent English mathematician Alan Turing 'Computing Machinery and Intelligence' published in 1950 in the October issue of *Mind* [18]. Hence, October of 1950 is the second (and, of course, the first in historical terms) date of commencement of research on AI. In his article, A. Turing formulated his famous test according to which the computer demonstrates an intelligent behaviour if it is capable of operating in such a way that an observer is unable to make out whom he is dealing with, a computer or a human being.

About 35–40 years have passed since then. the research on AI turned into a rather ramified scientific and technological branch encompassing a great deal of projects, hypotheses, feasible technological gains, etc. One is safe in saying that quite a few distinct advances have been made in AI, notably, in developing chess programs, in computer teaching, in expert systems design, development of military technologies (it is not for nothing that the military lavishly fund the AI research projects in many countries). Whatever the assessment of these achievements may be (a considerable scepticism is frequently displayed to this end, particularly by the natural intelligence experts), there is an indisputable fact: the sophisticated intelligent computer programs are capable of demonstrating an intellect- like behaviour in the local fields they are designed to function in. As applied to those fields, they meet fully the Turing test. And were it possible to reverse the history then, I believe, that Leibniz, who pondered much over the possibility of creating *thinking machines*, and A. Turing, who did not live until the Dartmouth Conference of 1956 (he died in 1954) would have been both satisfied with the current state-of-the-art in the AI field of research.

We, the people approaching the turn of the second millennium, are not, however, inclined to share such optimism. About 15 years ago A. Andrew, an English researcher, who has done much for realizing and popularizing the AI ideas, advanced an assumption that it would be possible to carry though the Turing test in its nontrivial form (i.e. with no constraints on the subject matter or response method) by 2000. Nevertheless, he went on, the computer will still resemble a rather constrained person with whom it is impossible to draw into a very modest talk. It is only later—believed A. Andrew—somewhere near 2050, that the computer might become a 'somewhat' with whom it would be possible to exchange a joke. And only then will the 'interlocutor' of a human find its own self to such an extent that people will have to think twice as to how to behave with it (see [2]).

I shall not guess as to what may happen by 2050—this is sooner an area of science fiction than of a rigorous prediction. As for the quite near 2000, there are no grounds for radiant perspectives. An up-to-date computer thinks no more than at the times of Turing, if thinking is defined as a process of creative solution of problems realized by the agent of thinking. Everything a computer is capable of doing is built in its programs by man, i.e. a programmer. Not a bit can the computer deviate from the program. Any unprogrammed situation, simple as it may be, turns into an insurmountable barrier for the computer. Computer self-learning is highly doubtful. Thus far, there have been no ways of passing the Turing test in its nontrivial form either.

Under the circumstances, a powerful boost to AI studies may apparently be given by a new and, hopefully, deeper insight into its foundations as a standard interdisciplinary area of science and technology. The interdisciplinariness is already built in the very name of the area. Artificial intelligence is, on the one hand, an intellect or mind, i.e. a sphere of interests of philosophy, logic, psychology, linguistics, physiology, etc. And, on the other, artificial, i.e. created by man with the technology available at each given instant in time, in this case—with the sophisticated computers. The historical paradox is that the interdisciplinary character of research on AI has always been implied but has never had an explicit impact on the strategy and progress of this kind of studies. At best, only some components of the complex interrelated AI domain were taken into consideration (thus, expert systems have to represent human knowledge and skills in computer programs). Time has come to represent the AI domain in an explicit and detailed form which may serve as a foundation for defining the more promising strategies of AI research.

Naturally, any AI conception should rely on the history and state-of-the-art AI research. True, there is no logic dependence of the AI conception on the relevant data of the AI history and modern state, but no AI conception can be deemed possible without due consideration of the findings of historical and contemporary research. It is also natural that vision and interpretation of the AI history and modern developments may bear a rather subjective character. Therefore, it would be naive to suppose that a single holistic conception of AI be possible at all. We have the right to speak of only our version of such a holistic conception, admitting at the same time that there can be other, alternative versions. The only thing that seems obvious here is the need for maximum consideration of at least the basic components of the AI holistic problem domain. In our version of the AI holistic conception such basic components are the following six interrelated blocks:

1. functioning of natural intelligence (NI);

2. development and implementation of intelligent computer programs;

3. natural intelligence theory;

4. intelligent programming theory;

5. systems-cybernetic paradigm;

6. philosophy, methodology and logic of knowledge, first of all scientific knowledge.

Taking into account the existing interrelationships between the blocks, the AI problem domain may be represented by the following diagram (the figures 1–6 signify the abovesaid blocks):

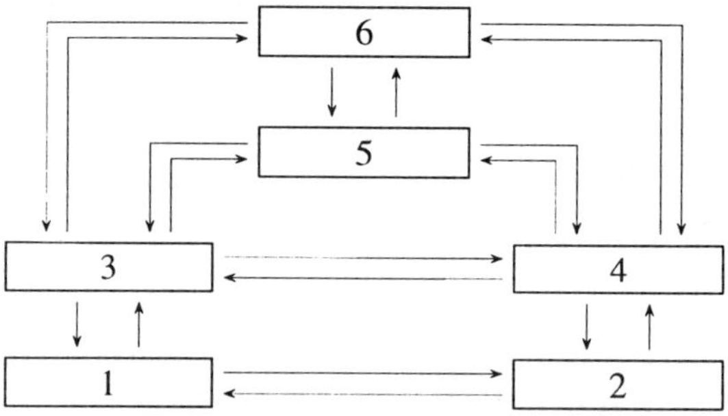

Let us first analyse the diagram from a historical point of view and then— in the context of the AI modern research. The initial AI problem setting implied linkage between Block 1 (NI functioning) and Block 2 (development and implementation of intelligent computer programs). It was supposed that if the AI project was feasible, then block 2 should act as a model of Block 1, and in a stronger version—as a real embodiment on another substratum of all the essential parameters of Block 1.

That historically initial version of AI did not deny the existence of an interrelationship between 1 and 2, on the one hand, and 3 and 4 on the other. The interrelationship between 2 and 4 (i.e. theory and practice of intelligent programming) was considered self-evident (in fact, it determined the successes expected in 2), whereas the impact of 3 on 1, and the feedback of 1 to 3, though admitted, had insignificant practical influence on 1 and even less—on 2. Naturally, the effectiveness of the NI practical functioning is determined by its improvement level that depends on how much the NI specifics is comprehended theoretically. Everywhere the assumption was (and it was generally quite reasonable) that our theoretical perception of the NI mechanisms is so fragmentary and uncertain that the available scarce knowledge is not capable of improving the NI operation, and even less—to serve as regulative guidelines for elaborating intelligent computer programs. As to more general and more abstract levels of description and substantiation of NI operation, in particular, Block 6— philosophy, methodology and logic of thinking, practically, the interrelationship of 6 and 1 (through, for instance, 3) was completely ignored, though no-one denied its existence.

Many authors viewed the connection between blocks 5 and 2 through block 4 as very important in the initial version of AI. Indeed, research on AI was born within the framework of cybernetic (more broadly, systems- cybernetic) paradigm. Therefore the impact of the general principles of cybernetics, general systems theory and systems approach on the theoretical perceptions of intelligent programming and, as a result, on practical implementation of intelligent programs was considered to be sufficiently obvious. However, this impact consisted more in declaration than in practical contribution to modifying 4 and 2.

Hence one can state that in the earlier period of the AI research, the real problem domain of AI reduced in fact to blocks 4 and 2 and their interrelationship. The existence of the remaining blocks, as we have pointed out, was recognised but their impact on the setting of the AI research problems was minor. This view on the AI problematique resulted in regarding AI as intelligent program systems autonomous of human knowledge (see, for example, [15]). The AI conception as a collection of autonomous intelligent computer programs reigned supreme in the AI field until the mid-1970s.

It was replaced by a program of expert systems construction, which introduced important modifications in the AI real domain. The new AI conception based on expert systems and so far prevalent in the field, made an impact on all the blocks of the AI problem domain but to differing degrees. Blocks 5 and 6 and their corresponding relationships with other blocks remained in the provinces of research interests. Blocks 4 and 2 with their interrelationship still held their important constructive role in identifying computer-related modes of knowledge formalization. Modified was only the natural enrichment of Block 4 content—first of all, by the development of knowledge representation methodology. The change in the AI new conception consisted of a totally different (from the earlier stage) understanding of the relationship between Blocks 1 and 2. Naturally, it involved a more profound content of Blocks 1 and 3 and a wider perception of the interrelationship between these elements of AI domain.

The principal distinguishing characteristic of the AI expert conception is that now it is suggested for feeding into the computer a quite definite 'mental material' (in the form stipulated by the content of blocks 2 and 4), notably, knowledge of experts in the given local subject area. Thus, at present the AI-related research starts with elicitation of expert knowledge (which is a far from easy job) followed by its analysis and 'improvement' (elimination of inconsistencies and incompleteness, etc.) and ends up as its formal computer representation. The resulting AI expert systems do not exist separately from man but serve, while interacting with users, as a considerable aid to man in decision making in the given local subject area (see [5, 7]).

Thus, one is safe in saying that the transition from the original to expert conception of AI expands the AI domain due to an explicit account of the relationship between 1 and 2. However, the new domain has not covered all of the essential components of AI either, therefore it could not be treated as something integral—either in theoretical realization or practical research on AI.

The current studies into AI problems employ a wide variety of specific research approaches. Probably, they have two common attributes: first, each approach focuses on a portion of blocks of the holistic representation of AI domain and, second, while other domain blocks are as if in the background, their impact on the studies into AI problems is ordinarily implicit—through an uncritical perception of some common sense, etc.

Thus, in fact, the '*purely computer people*' domain is confined by blocks 4 and 2 and relationships thereof. Here, the major emphasis is placed on the content of block 2, primarily concerning development of sophisticated methods of knowledge representation (frame approach and the like). The declared refusal to take account of blocks 1 and 3, let alone 6, is consistent with the research attitudes of '*purely computer people*', though it should not be taken at face value or literally. As a matter of fact, the ideas about the laws of the natural thinking process (of course, to the extent of their knowledge) influence both the structure of developed computer programs, and especially by the interpretation of the obtained results.

The proponents of other modern approaches to AI keep an eye in the final count, on block 1, i.e. on the real operations of natural intellect. Note that the approaches vary widely at that. Given a logical orientation, the focus is on the development of unconventional rules of logical inference, for example, rules of nonmonotonic reasoning (see, for instance [9]). One should not infer from this, however, that the latter is exactly the description of a natural intellectual creative process but its relation with such a process is not only recognized but is also the principal stimulus for the respective research.

The expert conception of AI, considered earlier, recognizes a much greater dependence of AI construction results on what we can elicit from the expert concerning the rules of intellectual reasoning he uses.

In other cases, the focus on block 1 results in a quite specific perception of the essence of all studies of this kind. Thus, for example, R. Schank and L. Hunter believe that probably the greatest significance of AI is not writing of new programs but rather the development of a new approach to understanding the process of thinking (see [13]).

A great deal of researchers treating AI from the standpoint of psychology of thinking, cognitive science and the analysis of laws of natural language functioning are rather sceptical with respect to AI potential capacities. The point of departure, in this case, is block 1, and in a wider context—the interrelationships of blocks 1 and 2 frequently reinforced by the respective philosophical and methodological concepts (block 6). The counter- arguments against AI—at least against a sound conception of AI (creation of an artificial intelligence, etc.)—are frequently rather convincing, but there is a strong feeling that far from always a due account is taken of the real capacities built in blocks 2 and 4.

Finally, the current research on AI is increasingly notable for a distinct approach to it from the standpoint of philosophy, methodology, logic (block 6) frequently in combination with the new concepts of systems- cybernetic paradigm (block 5). The consensus seems rather far away, however—the positions vary from some additional arguments in favour of psycho-linguistic criticism of AI to identification of philosophical and epistemological models of intelligence underlying and largely determining the feasibility of AI construction as it is realised today, primarily within the frameworks of blocks 2 and 4.

The painted picture of the current state-of-the-art in AI research seems to be a sound argument justifying the necessity of designing a holistic conception of AI problems.

2 How is Natural Intelligence Possible?

The present and subsequent sections of the paper will deal with a number of important (from my standpoint) AI problems proceeding from the general characteristic of the holistic approach to AI considered in Section 1.

The first of such problems is: *how is natural intelligence possible*? This question seems senseless at the empirical level: the entire history of humanity is an illustration of reality, hence—the possibility of natural intelligence (NI). In the Kantian sense, however, the question of theoretical possibility, i.e. theoretical justification of NI is quite lawful. What is more, in spite of the fact that in the course of human knowledge evolution this question was frequently raised—in some form or other—there is still no generally acceptable answer to it.

Characteristic to our approach to the question as to how NI is possible is that we shall treat this question not only and not so much within the frameworks of NI theory (blocks 3 and 1 in the diagram from Section 1), but with due regard to NI and AI relationships (blocks 1 and 2), and in general in the context of the entire holistic model of AI, i.e. by making use, in particular, of the content of blocks 5 and 6.

As is well known, the term 'intelligence' or 'mind' is rather polysemantic: it has a protracted history, its meaning has frequently varied considerably following alternation of philosophical traditions. At the same time, given all of the meaning modifications, which we are unable to analyse here, intelligence, on the one hand, is associated with the capacity of thinking for rational cognition and, on the other, is contrasted to the spiritual capacities such as feeling, will, imagination, and the like. It is within the above two semantic determinants, i.e. rationality and sensual cognition, that one should look for an answer to the question as to how natural intelligence is possible.

Guided by the notions, principles and concepts contained in the modern versions of blocks 3, 5, and 6, we shall identify a number of essential characteristics of intelligence or mind:

- Intelligence is first of all a certain form of spiritual activity, possessing its own subject, its elements, means and standards. It also realizes certain goals.
- The subject of natural intelligence is man or mankind as a whole.
- By elements of intellectual activity we understand the basic elements—notions, conceptions, judgements, statements, etc., expressed materially in various languages—natural and artificial.
- Intellectual activity represents unity of syntactic (formal) and semantic (meaningful) transformations of the activity elements expressed in a linguistic, verbalised form.
- Intellectual activity is rational, i.e. each of its steps should be substantiated by means of the earlier adopted standards (criteria) of such activity.

- The purpose of intellectual activity is to state and solve the tasks or problems facing the subject.
- Being but a part or aspect of man's spiritual activity, intellectual activity is 'surrounded' by forms of sensual knowledge and is performed, as a rule, in a complex interaction with the last. Only by abstracting our thinking process, are we able to separate the intellectual activity proper, on the one hand, and sensual knowledge, on the other.

The abovesaid essential characteristics of intelligence or mind require explanation and specification. It is, in the first place, necessary to classify the collection of intellectual problems into two categories—*creative problems* (when set, neither humanity nor the given individual have a method of their solution) and, so to speak, *routine* or *trivial problems* (mankind is familiar with their solutions but each time they have to be looked for as if anew).

The two classes of the intellectual problems identified may be interpreted differently: when trivial problems are set, all the tools needed for their solution (elements, rules, criteria and so on) are available. With creative problems, however, at least some components of their solution are missing, and they have to be discovered or invented. Naturally, in the course of the historical development of natural intelligence, the creative problems gradually become trivial ones.

It is necessary to explain here in what sense, from our viewpoint, one should understand the assertion of the rational character of intellectual activity. For the last two or three decades, the notion of rationality caused a lot of 'battles' but they failed to result in a generally adopted view. In the course of the discussions, however, their participants succeeded in realising one important point: the rational is broader than the logical. As was stated by B. S. Gryaznov,

> "the modern logical theories do not cover the entire domain of the rational ... even in the remote future there will not be complete identification of the rational and the logical, as it contradicts the spirit of the most up-to-date logical theories" [6, p. 211].

I fully share this opinion.

Thus, if one subtracts the logical from the rational, there may remain something substantial. How is one to understand it? Here, I think, we may rely on the knowledge of the nature of the logical, where one can clearly single out two aspects: adherence to the earlier formulated rules (criteria, rules of deduction) and the conservation of validity of the results obtained (given the truth of the premises, the logical inference inevitably leads to a true conclusion). The rational, which is not logical, does not possess the quality of validity conservation but meets the earlier formulated criteria. This adherence to the clear-cut rules and criteria makes it possible to substantiate each step of the rational reasoning. As a result, the rational is synonymous to substantiation on the basis of the earlier adopted rules.

It is obvious that the routine intellectual operations fully satisfy the abovesaid conception of the rational. As to creative thinking, we have already stated that there are no general solutions to the problems at the moment of their formulation. Therefore, in many cases, their solution depends on the formulation of new rules and criteria. There is an impression that creative intellectual operations go beyond the suggested conception of the rational. However, it is not so.

In fact, the creative intellectual process involves the construction of a new intelligent system meeting all rationality requirements. The transition from the old intelligent system to a new one seems irrational only against the criteria of the previous system. Once the new system is built and its rationality criteria formulated, it is again viewed as a rational structure. Hence, the evolvement of natural intelligence is precisely the process of a subsequent transition from one rational structure to other rational structures.

We are now in a position to answer the question put in the beginning of this section. Natural intelligence is possible because, while being a certain rational structure at each instance of its existence and faced with the necessity to solve innovative problems, it acts as an irrational activity from the standpoint of old rationality criteria, but as soon as there is a solution to those problems the natural intelligence again turns to be a rational structure.

3 Can Syntax Generate Semantics?

The above answer to the question 'How is natural intelligence possible?' holds equally for the possibility of NI and AI. This follows, in particular, from our research methodology operating on the holistic conception of AI which focuses not so much on NI or AI, but intelligence or mind in general.

At the same time the answer given in Section 2 as applied to AI, generates a serious epistemological difficulty. Indeed, the sophisticated computer is doomed to function only in line with the instructions designed by the programmer in a computer program. There are no feasible ways out of the situation, at least in the foreseeable future. As a result, the computer handles perfectly routine intellectual processes (sometimes it is superior to the human being) but it is absolutely incapable of finding a solution to the problem not anticipated in advance. Therefore, any problem the computer is unable to solve at any given instant in time, is considered creative, in fact a true intellectual problem. However, once the method of its solution is built in the program, it turns into a routine problem placing demand on just a multitude of mechanical operations. Artificial intelligence happens to be artificial but not intelligent. This self-inconsistency of AI project was clearly fixed, for example, by A. Andrew, who pointed out that any attempt to create an intelligent or thinking machine automatically implies self-negation. As long as the development of artificial intelligence does not accomplish its higher goal—duplication of each aspect of human intelligence (which is, apparently, unfeasible in the foreseeable future) such self-negation will, no doubt, be characteristic of it (see [3]).

The task of duplicating each note, each aspect of human mind is absolutely unrealistic and, what is more, such a goal is just senseless. And if AI happens to be comparable with NI with respect to its capacities, then it makes sense to reproduce something more perfect than human NI which, for all the geniality of individual creators, is always known to have distinct limitations. Should humanity be ever capable of solving this problem, even then the developed AI will not duplicate each aspect of human intellect.

As for the current state-of-the-art in the area of AI then, as was noted earlier, in many local areas AI satisfies the Turing test. This means that the epistemological difficulty, formulated in the beginning of this Section, can be resolved as follows.

The argument on self-inconsistency or self-negation of AI development problem should by no means be interpreted absolutely. It holds with respect to the process of devising solutions of creative intellectual problems. According to the definition of intelligence or mind, accepted in this paper, a thinking agent must, in principle, find ways of solving creative problems. Man is capable of doing it. The computer does it if only assisted by man, and in this respect the considered argument is valid. One should not infer from this, however, that a computer is unable to exercise an intelligent activity, particularly with regard to the fact that while evolving computer programs have increasingly extended the ever new intellectual spheres of human activity. And it does not follow from this argument at all that the very statement of the AI creation problem is inconsistent. The argument just specifies the idea behind our understanding of AI construction problem.

We may now turn directly to the question formulated in the heading of the present section—'can syntax generate semantics?'. A sharply negative answer to this question was given quite recently by John Searle of the University of California at Berkeley [16]. The latter, however, does not substantiate his answer, he just treats it as an axiom.

Indeed, Searle's article discusses the question as to whether the human mind or intelligence—the brain activity product—is a computer program. Searle tries to substantiate his answer. He operates on the following three axioms:

Axiom 1. Computer programs are formal (syntactic) systems.

Axiom 2. Human minds have mental content (semantics).

Axiom 3. Syntax, by itself, is neither constitutive of nor sufficient for semantics.

from which he concludes:

Conclusion 1: Programs are neither constitutive of nor sufficient for minds.

By adding yet another axiom:

Axiom 4. Brains cause minds,

Searle arrives, after all, at his final conclusion:

Conclusion 4: The way that the human brain actually produces mental phenomena cannot be solely by virtue of the running of a computer program [16, p. 21–23].

The space of this article does not allow a detailed analysis of Searle's reasoning. I would like to draw your attention to just several points relevant to the topic of the paper. First, I believe that the final conclusion of Searle, notably—Conclusion 4 is indisputable, but it is not identical to his earlier Conclusion 1. The recognition that there is something more than a program in intelligence is much more inferior to the proposition that whatever programs the computer implements, it does not follow from it that it has intelligence. Second, I do not consider the first Searle conclusion correct either, from the logical point of view, primarily due to equivocality of his third axiom.

In formulating that axiom, Searle proceeded from the traditional contrasting of syntactic transformations, abstracting from any meaningful components, to semantics, i.e. meaningful transformations. Of course, given this approach, syntax does not convert, by definition, into semantics.

At the same time, any syntactic system is an ordered set of different syntactical units: signs, formula, rules, operations, etc. Put together, they fragment the operation universe of each syntactic system, into various syntactically connected 'blocks'. Indeed, there are no semantics here—at least in explicit form, there is room, however, for semantics—so to speak, potential semantics. And once we interpret the syntax components in semantic terms, then all meaningful characteristics representable in the given syntax will become explicitly distinct.

The man acts in such a way that the syntax and semantics come to be closely interconnected. The computer is incapable of 'pondering' over its operations. But performing in a purely syntactical sphere it designs some syntactic frameworks of potential semantic interpretations, and once somebody exercises such interpretation, the syntactic inferences of the computer happen to be meaningful.

The above reasoning testifies to a deep, organic parallelism of syntax and semantics [1]. It follows from here that only syntactic transformations are possible in the syntactic sphere, meaning no syntactical equivalents, but this automatically generates syntactic schemes for the respective semantics. As a result, we may give the following answer to the question put in the heading of the current Section.

True a syntax, deprived of semantics by definition, is unable to generate it. And Searle is right in this respect. But each syntactic system is a potential scheme of a variety of potential semantics, and should one interpret 'to generate' as to create conditions for the potential being, then each syntax automatically generates semantics. And this proposition questions the truth of the third Searle axiom.

It is worth noting that the answer to the question was obtained within the framework of a holistic concept of the AI domain. Indeed, in articulating the answer we made use of not only the data from blocks 1,2, 3, and 4 from our scheme from Section 1, indicative of the distinguishing characteristics of NI and AI, but also philosophical, logical, and systems-cybernetic concepts of intellect or mind contained in blocks 5 and 6.

[1] Note that the idea of such parallelism—true, in a specific terminology characteristic of Soviet philosophical literature of the 1950s–1960s—was advanced by G. P. Shchedrovitsky in 'the principle of parallelism of the form and content of thinking' and formulated in the mid 1950s [14].

Now, in the closing section of the paper, let us try to outline, proceeding from a holistic concept of AI, some prospective lines of AI research.

4 Some Immediate and Distant Prospects of AI

The future of research on AI is in the focus of attention today. We can hardly resist the temptation too, the more so that we hope that the suggested holistic concept of AI can be helpful not only in the practical implementation of AI projects but also will open up new avenues in the discussion of AI prospects. Therefore, let us again turn to our scheme from Section 1 and start the discussion with block 6 and its possible impact on other blocks.

It is indisputable for a philosopher that any conceptual research into AI problems makes use—explicitly or implicitly, consciously or unconsciously—of some or other philosophical concepts and models of intellect, mind, thinking, creative thinking, etc. Recently, this philosophy behind the AI came to be the subject of an explicit analysis (see a number of papers by D. C. Dennett, H. Putnam, D. Davidson, H. Melloz, and many others; see also [4, 17]). The two relevant inferences merit, I believe, a serious attention.

First, the relationship (identified by a number of researchers) between the modern concepts of intelligence as a whole and AI, in particular, and the classical concepts of philosophical rationalism advocating fragmentation of the object under study into its primary parts; through the analysis thereof, determination of their relationships, and so on, followed by attempts to restore the whole with such primary blocks. An old, good tradition but not without essential limitations: it is impossible to restore the whole going this way, at least for rather complex cases. Hence, one should not overestimate its capacities as applied to AI. It is very important to clearly define these capacities today.

Second, there is an increasingly distinct trend to approach intelligence from a systems, holistic perspective as an alternative to classical rationalism. In this case, the construction of AI starts as if in reverse direction—not with its components but with the whole. The potential of such an approach is huge, but the ways of its implementation are rather vague. Of considerable interest in this connection is a study conducted by K. L. Viktorova, who managed to formulate the principles of a moderate holism based on the ideas of the late L. von Wittgenstein as a philosophical basis for analysing AI problems (see [19]).

D. A. Pospelov, a prominent Russian scholar, approaches the future AI research from a somewhat different perspective. He also proceeds from the impact of blocks 6, 5, and 3 on the current AI studies, and suggests switching over from the logical AI paradigm, valid only within formal systems, to a paradigm of argumentation, and then to the paradigm of justification of the fact within the frameworks of the current values (see [11]). Of course, the point here is about the distant perspectives—rather appealing but hardly feasible with regard to the present-day content of blocks 2 and 4. Nevertheless, the search along these lines constitutes a real way of exercising a holistic approach to AI.

The discussion of development prospects in any field of science is always open and incomplete. We do not claim a full coverage of the topic either, but will just mention a handful of important areas of the further studies into AI problems:

- Construction of formal logical systems most suitable for the description of a natural process of reasoning, in particular the development of nonmonotonic logics (J. McCarthy, V. Lifschitz, *et al.*, for example [9] and [8]).
- Development of mathematical techniques of intelligent programming related, in particular, to problems of data bank construction, knowledge representation, and so on. (G. M. Adelson-Velsky, V. L. Arlazarov, M. V. Donskoy, V. V. Yurchenko and many others, see for example [1] and [4]).
- Implementation of alternative computer architectures exercising parallel distributed processing, neuron networks, etc. [12].
- Studies into common sense, mechanisms of natural language usage, etc. (T. Winograd, I. Hacking, R. Harre, *et al.* see, for example [20]).

And, finally

- Development of methods of an intelligent interface between computer and man (see, for instance [10]). Computers will not be able to function independently of man in the forthcoming future, therefore a deeper insight into their interaction is one of the major objectives of AI research projects.

Naturally, the aforementioned lines of research into AI problems are, so to speak, 'built-in' the holistic research domain of AI and the real gains could be expected within the frameworks of this kind of the holistic approach to AI.

Russian Academy of Sciences, Moscow

References

[1] G. M. Adelson-Velsky, V. L. Arlazarov, and M. V. Donskoy. *Algorithms for Games*, Springer-Verlag, New York, 1988.

[2] A. M. Andrew. Possibilities and probabilities. *The Robots are Coming*, F. H. George and J. D. Humphries, eds. Ch. 8, pp. 115–123, Manchester National Computing Centre, 1974.

[3] A. M. Andrew. *Artificial Intelligence*. Abacus Press, 1983.

[4] V. N. Sadovsky and V. N. Kostyuk, eds. *Artificial Intelligence and Problems of Knowledge Organisation. VNIISI Collected Papers*, Moscow, 1991. (In Russian.)

[5] V. A. Gelovani, O. V. Kovrigin and N. D. Smolianinov. Methodological problems of building expert intelligent systems. *Systems Research. Methodological Problems Yearbook*, pp. 254– 278, Nauka Publishers, Moscow, 1983. (In Russian.)

[6] B. S. Gryaznov. *Logic, Rationality, Creativity*. Nauka Publishers, Moscow, 1982. (In Russian.)

[7] D. I. Larichev, A. I. Mechitov, H. M. Moshkovich and E. M. Furems. *Elicitation of Expert Knowledge*, Nauka Publishers, Moscow, 1989. (In Russian.)

[8] *Logic and Computer. Modelling of Reasoning and Program Verification*. Nauka Publishers, Moscow, 1990. (In Russian.)

[9] J. McCarthy. Mathematical logic in artificial intelligence. *The Artificial Intelligence Debate. False Starts, Real Foundations*, St. R. Graubard, ed. pp. 297–311, MIT Press, Cambridge, MA, 1988.

[10] M. S. Elzas, T. I. Oren and B. P. Zeigler, eds. *Modelling and Simulation Methodology. Knowledge Systems Paradigms*. North-Holland, Amsterdam, 1989.

[11] D. A. Pospelov. From inference to justification: a change over of paradigms in artificial intelligence. *Formal and Informal Reasonings*, Transactions of a Seminar on Artificial Intelligence, pp. 74–81, Tartu University Collected Papers, 1989. (In Russian.)

[12] D. F. Rummelhart, and J. L. McClell and the PDP Research Group, eds. *Parallel Distributed Processing: Explorations in the Microstructure of Cognition*. Vol 1 and 2. MIT Press, Cambridge, MA, 1986.

[13] R. Schank and L. Hunter. The quest to understand thinking. *Byte*, April 1985.

[14] G. P. Shchedrovitsky. Language thinking and methods of its analysis. *Voprosy yazykoznaniya*, 1957, N1. (In Russian.)

[15] Yu. A. Schreider. Computer as a means for knowledge representation. *Priroda*, 1986, N10,pp. 14–22. (In Russian.)

[16] J. R. Searle. Is the brain's mind a computer program? *Scientific American*, **262**, 20–25, 1990.

[17] St. R. Graubard, ed. *The Artificial Intelligence Debate. False Starts, Real Foundations*, MIT Press, Cambridge, MA, 1988.

[18] A. M. Turing. Computing machinery and intelligence. *Mind*, **59**, 433–460, 1950.

[19] K. L. Viktorova. Specific utilisation of theoretico-cognitive constructs in artificial intelligence systems. In [4]. Moscow, 1991. (In Russian.)

[20] T. Winograd. Artificial intelligence and language comprehension. *Artificial Intelligence and Language Comprehension*, Washington, DC, 1976.

Part II

Modern Logic and Methodology of Science

4

Hierarchical Learning of Boolean Functions

Patrick Suppes and Shuzo Takahashi

1 Introduction

The learning of Boolean functions has been the focus of a number of papers in computer science in the last few years, many of which have been stimulated by the work of Valiant [3, 4]. We also mention especially the recent long article by Blumer *et al.* [1] on learnability for various classes of concepts including especially Boolean classes. The thrust of most of this theoretical work has essentially been to show under what conditions concepts can be learned in polynomial time or space. It is of course important to show that learning is feasible for as wide a class of concepts as possible. However, it is also important to focus on detailed results of a very finitistic kind, which are characteristic of learning theory as applied to human behavior, and are of intrinsic interest also in machine learning. In this paper we consider conditions under which a hierarchy of Boolean functions can increase in significant ways the rate of learning of a given Boolean function of n variables. In this context we also compare serial and parallel hierarchical learning, with, as might be expected, the most dramatic effects being produced by parallel hierarchical learning. Because in general the results are better for parallel than for serial learning the reader might wonder why we include a systematic study of serial learning. The reason is that even under serial learning conditions hierarchical learning can often lead to dramatic improvement in the rate of learning.

There is some literature on hierarchical learning, but the results are not closely related to ours. We refer the reader to Rivest and Sloan [2], and the references given there.

There are three distinct psychological processes involved in our theoretical approach to hierarchical learning. They are (i) stimulus sampling, often given in the form of a pattern; (ii) concept sampling; and (iii) function formation—meaning forming some function of the concepts sampled. In some contexts it may be useful to refer to function formation as hypothesis sampling. In any case, process (i) brings in information from the world outside, but (ii) and (iii) are essentially internal.

P. I. Bystrov and V. N. Sadovsky (eds.), Philosophical Logic and Logical Philosophy, 51–61.

With probabilistic assumptions about each process, a variety of measures of learning can be defined and many of them have been used in the literature. Because we are interested in comparing a number of different assumptions about each process in terms of effects on the rate of learning, we have chosen as the comparative measure we compute the expected number of trials to learn the concept in question, either with no residual response error in the model or with some specified lower bound on the probability of an error or the expected number of trials. By and large we do not analyze possible processes of function formation, in spite of their interest for a general theory of learning. We comment on this point in more detail later.

1.1 Relation to neural networks

It is important to emphasize that hierarchies, as defined and used in this article, are not neural networks, except in the most trivial sense. The reason is simple. In the hierarchies we consider there are no hidden units. Each level of a hierarchy can provide reinforcement or feedback on learning any Boolean function occurring at that level. This assumption is made for all the different cases we consider.

On the other hand, wide variations in sampling assumptions are made. The most important dichotomy is between (i) the assumption of sampling of input to a Boolean function at one level of a hierarchy being independent of such sampling at any other level, and (ii) the assumption that the sampling distribution is given at the lowest level of the hierarchy, the level of stimulus patterns, and this distribution then determines the sampling at all higher levels.

We underscore the point that it would be a mistake to suppose that sampling assumption (ii) is almost always more realistic than assumption (i). In the teaching of a complex hierarchical subject like mathematics or a foreign language, students in general sample artificial exercises at each level of learning, essentially independent of their distribution of occurrence as components of real-world problems to be encountered later.

The hierarchies we consider are not neural nets, but it is obvious that a neural net could be added at each node by so formulating the learning of each Boolean function at a node of the hierarchy. This we have not done, for reasons that will become clear later.

2 Some Preliminaries

A Boolean function of n variables is a function from $\{0,1\}^n$ to $\{0,1\}$. We use variables F, G, H, F_1, F_2, etc. for Boolean functions. A pattern is an element in $\{0,1\}^n$. Variables p, q, p_1, p_2, etc. are used for patterns. When there are n variables, then there are 2^n patterns, and Boolean functions can also be defined by subsets of the 2^n patterns. For this reason we shall sometimes refer to Boolean functions as concepts defined by subsets of the patterns. The usual convention is that the

Boolean function has the value 1 for the patterns that are instances of the concept, and 0 otherwise.

The kind of learning process we study here is what is called in the literature supervised learning from examples only. This means in classical learning terms that the learner is only presented examples and is informed after each trial what is the correct response. We can assume that on each trial the learner guesses whether a given pattern is an instance of the concept or not, that is, whether the Boolean function has the value 1 or not for this pattern. Note that under the assumption of feedback we are making here it is not important whether the response is correct or incorrect. What is important is the assumption that the learner internalizes the correct answer.

We can now note a familiar result, but one that is important for comparison. If the learner is asked to learn an arbitrary Boolean function purely from presentation of examples, then in general there are for n variables at least 2^n trials required. To achieve this lower bound of 2^n, we must assume that the learner is perfect in the sense of remembering and correctly registering the feedback on each trial. Then there is no way that the learning can be speeded up without the introduction of information that goes beyond the pure presentation of examples. This means that for a function of 1000 variables, which could easily arise in a vision system, the learning of patterns would require 2^{1000} trials, obviously an unfeasibly large number. Valiant and others have emphasized learning restricted classes of functions that will not require such an exponentially large number of trials. The analysis that we give of hierarchical learning is similarly motivated but moves in a different direction.

We introduce some further notation useful in studying hierarchies. The variables x, y, z, x_1, x_2, etc. take values in $\{0, 1\}$, and f, g, h, f_1, f_2, etc. are function symbols with certain arities. $A(f)$ denotes the arity, or number of arguments, of f. Terms are recursively defined from variables and function symbols. For each term t, we define a set Q_t which contains all function symbols in t. This can be recursively defined (we omit the precise definition). For example, for $t = f(x, y)$, $Q_t = \{f\}$, and for $t = f(g(x_1, x_2), h(x_1, x_2, x_3))$, $Q_t = \{f, g, h\}$. Also $|Q_t|$ is the cardinality of Q_t. For each term t, the number of distinct variables in t is denoted by V_t. For example, for $t = f(x, y)$, $V_t = 2$, and for $t = f(g(x_1, x_2), h(x_1, x_2, x_3))$, $V_t = 3$.

Throughout we use, with or without arguments or subscripts, N for nonhierarchical learning, S for serial hierarchical learning and P for parallel hierarchical learning.

3 Serial Hierarchical Learning

In this section we compare nonhierarchical and serial hierarchical learning of Boolean functions. Intuitively the comparison of nonhierarchical and serial hierarchical learning in the framework we have defined is straightforward, given that in the present analysis we are concerned with perfect learning. Later we shall consider learning to a particular statistical criterion, of the sort discussed by Valiant. The nonhierarchical learning of an arbitrary Boolean function of n variables is just learning the correct response to each of the 2^n patterns. The hierarchical learning depends upon learning each function in a corresponding way. In other words, each individual function, at any level in a hierarchy, is a special case of nonhierarchical learning. More explicitly for each term t, N_t and S_t are defined by

$$N_t = 2^{V_t}$$

and

$$S_t = \sum_{f \in Q_t} 2^{A(f)}.$$

These two quantities are used to measure the rates for nonhierarchical and serial hierarchical learning, as just mentioned above. For example, for $t = (f(x,y), N_t = 2^2 = 4$ and $S_t = 2^2 = 4$, for $t = f(g(x_1,x_2), h(x_1,x_2,x_3)), N_t = 2^3 = 8$ and $S_t = 2^2+2^2+2^3 = 16$, and for $t = f(g(x_1,x_2,x_3), g(x_2,x_3,x_4)), N_t = 2^4 = 16$ and $S_t = 2^2 + 2^3 = 12$.

Note that the quantities N_t and S_t measure exactly the rates of learning under the assumptions that *sampling is without replacement and that at each level of a hierarchy only relevant concepts are sampled.* There is no need to say sampling is uniform. The only requirement is that on each trial a pattern is sampled that has not been sampled before. When studying perfect learning under the nonreplacement assumption, what actual distribution is used for sampling does not affect the number of trials for the results we are now computing.

The most extreme sort of example favoring serial hierarchical learning over nonhierarchical learning is given by:

$$t = h(g(f(x_1,x_2), f(x_3,x_4)), g(f(x_5,x_6), f(x_7,x_8))). \tag{1}$$

For this t,

$$N_t = 2^8 = 256,$$

while

$$S_t = 3 \times 2^2 = 12.$$

More generally, for any term t with n variables, which has a similar structure. $N_t = 2^n$ while $S_t = 4 \log_2 n$.

The most extreme example favoring nonhierarchical learning is given by:

$$t = h(g_1(f_1(x_1,x_2), f_2(x_1,x_2)), g_2(f_3(x_1,x_2), f_4(x_1,x_2))). \tag{2}$$

For this t,

$$N_t = 2^2 = 4,$$

while

$$S_t = 7 \times 2^2 = 28.$$

A less extreme yet interesting example, also of interest for parallel processing, is given by:

$$t = h(g_1(f_1(x_1, x_2), f_2(x_3, x_4)), g_2(f_3(x_5, x_6), f_4(x_7, x_8))). \quad (3)$$

For this t,

$$N_t = 2^8 = 256,$$

while

$$S_t = 7 \times 2^2 = 28.$$

Figure 1 shows the hierarchy of t as defined by (3). And more generally, for any term t with n variables, which has a similar structure, i.e., n is a power of 2, $N_t = 2^n$ while $S_t = 4(n-1)$.

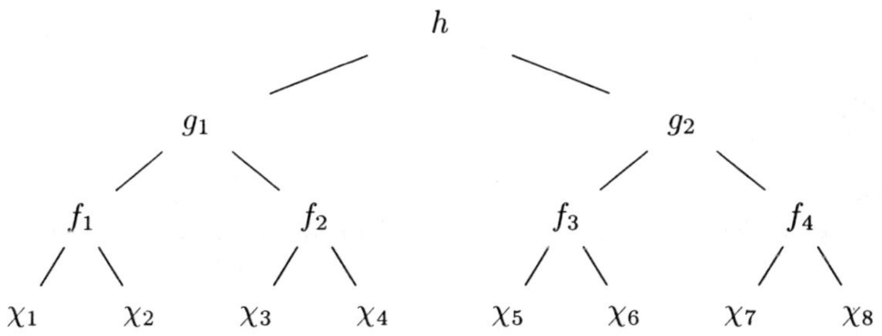

Figure 1: Hierarchy defined by Equation (3).

3.1 Sufficient conditions for $S \le N$

We now examine sufficient conditions for $S \le N$ (henceforth we omit the subscript t from the notation), that is, for

$$\sum_{f \in Q} 2^{A(f)} \le 2^V. \quad (4)$$

First of all, it is easy to show that

Theorem 3.1 *If* $V \le 3$, *then* $N \le S$.

Thus our main interest is in the case $V \ge 4$. It is difficult to analyze (4) without a further simplification[1]. We start with the following simple (rather crude) estimation of S as the basis of our analysis of $S \le N$:

$$\textstyle\sum_{f \in Q} 2^{A(f)} \le |Q| 2^M \quad (5)$$

[1] We might also consider some structural assumptions, such as the depth of terms is 1. However, it seems that such cases are already as difficult as the general case, without further simplifications as exemplified below.

where $M = \max\{A(f) : f \in Q\}$. Then, a sufficient condition for $S \leq N\ (= 2^V)$ is

$$|Q|2^M \leq 2^V. \tag{6}$$

Now two cases may be considered: $M = 2$ and M/V is constant, i.e., $M = hV$ for some constant h.

These two cases are further analyzed in the following two subsections.

3.1.1 Analysis of $S \leq N$ in the case $M = 2$

Putting $M = 2$ and $|Q|/V = k$ where $0 < k$, (6) becomes

$$4kV \leq 2^V. \tag{7}$$

It is easy to show that a necessary and sufficient condition for (7) (for $V \geq 4$) is $k \leq 1$. Thus we obtain

Theorem 3.2 *For a binary term with at least 4 variables, if the number of the function symbols in the term does not exceed the number of the variables in the term, then $S \leq N$.*

It is also interesting to consider the ratio $r = S/N$ in the case $S \leq N$. For this, replacing k by k/r in the above argument, we obtain

Theorem 3.3 *For a binary term with at least 4 variables, if $|Q|/V \leq r$, then $S/N \leq r$.*

3.1.2 Analysis of $S \leq N$ in the case $M = hV$

Putting $M/V = h$ and $|Q|/V = k$ where $0 < h < 1$ and $0 < k$, (6) becomes

$$kV \leq 2^{(1-h)V}. \tag{8}$$

Let us consider the function

$$\phi(V) = 2^{(1-h)V} - kV. \tag{9}$$

The function ϕ has its minimum at

$$m(h,k) = \frac{1}{(1-h)\log 2} \log \frac{k}{(1-h)\log 2}.$$

Hence, the inequality (8) holds for all V if and only if

$$\phi(m(h,k)) \geq 0,$$

which is equivalent to

$$k \leq (e \log 2)(1-h). \tag{10}$$

Thus we obtain

Theorem 3.4 *A sufficient condition for $S \leq N$ is that (10) holds. (Notice that if (10) holds for h and k, then it holds for any $h' \leq h$ and $k' \leq k$).*

We are also interested in the ratio $r = S/N$ in the case $S \leq N$. This can be easily obtained by replacing k by k/r in the above derivation, that is,

Theorem 3.5 *If the following inequality holds*

$$\tfrac{k}{r} \leq (e \log 2)(1-h), \tag{11}$$

then $S/N \leq r$.

Note that $e \log 2$ is about (yet bigger than) 1.88. As an instance of (11), if $M/V \leq 0.60$ and $|Q|/V \leq 0.56$, then $S/N \leq 0.75$.

In spite of starting with the rather crude estimation (5), the numbers we obtained for h, k, and r look reasonable. Now let us consider what kind of improvements for h, k, and r can be made. First of all, notice that Theorems 3.4 and 3.5 are valid for all V, and do not reflect the following fact:

If V gets larger, then (8) holds for larger h and k.

To use this fact, let us consider the following condition:

$$C(h,k,V_0) \ \textit{if and only if for all} \ V \geq V_0, kV \leq 2^{(1-h)V}. \tag{12}$$

Note that if we want to consider the ratio $r = S/N$, then simply replace k by k/r. Simple properties of this condition are:

- For all h and k, there exists V_0 such that $C(h,k,V_0)$.
- If $C(h,k,V_0)$, $h' \leq h$, and $k' \leq k$, then $C(h',k',V_0)$.
- If we expect V_0 to be small, then both h and k must also be sufficiently small.
- If we expect h and k to be large, then V_0 must also be sufficiently large.
- Even though h is large, if k is sufficiently small, then V_0 can be reasonably small.
- Even though k is large, if h is sufficiently small, then V_0 can be reasonably small.

Next we pose the problem: Given V_0, find a simple relation between h and k such that $C(h,k,V_0)$.

Theorems 3.4 and 3.5 were obtained by considering $\phi(m(h,k)) \geq 0$, but this does not give a good result in the case $m(h,k) \leq V_0$, which is equivalent to

$$k \leq ((1-h)\log 2) 2^{(1-h)V_0}. \tag{13}$$

In this case, a sufficient condition for $C(h,k,V_0)$ is $\phi(V_0) \geq 0$, which is equivalent to

$$k \leq \frac{1}{V_0} 2^{(1-h)V_0}. \tag{14}$$

Combining (13) and (14), we obtain

Theorem 3.6 *If*

$$k \leq \min\{(1-h)\log 2, \frac{1}{V_0}\} 2^{(1-h)V_0},$$

then for $V \geq V_0$

$$S \leq N.$$

Substituting k/r for k, we obtain

Theorem 3.7 *If*

$$k/r \leq \min\{(1-h)\log 2, \frac{1}{V_0}\} 2^{(1-h)V_0},$$

then for $V \geq V_0$

$$\frac{S}{N} \leq r.$$

3.2 Uniform Sampling with replacement of stimulus patterns

Let E_N be the expected number of trials to sample all stimulus patterns under non-hierarchical learning when the sampling distribution is uniform with replacement, and let E_S be the corresponding expected number for serial learning with the sampling distribution uniform with replacement for each function at any level of the hierarchy. Thus, for example, if we consider the hierarchy of Figure 1, the expected number of trials to sample all patterns for each of the 7 binary functions is 25/3, so $E_S = 58.3$. These numbers follow at once from the following familiar results.

Let E_N be the expected number of trials under the uniform distribution to completely sample with replacement all N patterns. First, the sampling may be represented by the Markov chain whose transition matrix is given by:

$$\begin{array}{c} \\ 0 \\ 1 \\ j \\ N-1 \\ N \end{array} \left(\begin{array}{ccccccc} 0 & 1 & 2 & j & j+1 & N-1 & N \\ & 1 & & & & & \\ & \frac{1}{N} & \frac{N-1}{N} & & & & \\ & & & \frac{j}{N} & \frac{N-j}{N} & & \\ & & & & & \frac{N-1}{N} & \frac{1}{N} \\ & & & & & & 1 \end{array} \right)$$

The probability distribution of the number of trials in the state $j < N$ is given by:

$$P_{N,j}(n) = \left(\frac{j}{N}\right)^{n-1} \left(1 - \frac{j}{N}\right).$$

Hence,

$$E_{N,j} = \sum_{n=1}^{\infty} nP_{N,j}(n) = \sum_{n=1}^{\infty} n \left(\frac{j}{N}\right)^{n-1} \left(1 - \frac{j}{N}\right) = \frac{1}{1 - \frac{j}{N}} = \frac{N}{N - j}.$$

So, the expected number of trials for completely sampling all N patterns is:

$$E_N = \left(\sum_{j=1}^{N-1} E_j\right) + 1 = \left(\sum_{j=1}^{N-1} \frac{N}{N-j}\right) + 1 = N \sum_{k=1}^{N} \frac{1}{k}.$$

For $N = 4$,

$$E_N = 4(1 + \frac{1}{2} + \frac{1}{3} + \frac{1}{4}) = \frac{25}{3},$$

as stated above. (Note that learning for any given function in the hierarchy is itself nonhierarchical, just depending on the number of patterns to be sampled.) In contrast to $E_S = 58.3$ for Figure 1, the nonhierarchical learning of 2^8 patterns under uniform sampling is

$$E_N(8) = 256 \sum_{k=1}^{256} \frac{1}{k},$$

which is a much larger number. Explicitly,

$$E_N(8) \approx 1567.3$$

More generally, for the same hierarchical structure of binary functions as that of Figure 1, but for arbitrary n,

$$E_S(n) = \frac{25}{3}(n - 1),$$

for uniform sampling without replacement, and for the nonhierarchical case

$$E_N(n) = 2^n \sum_{k=1}^{2^n} \frac{1}{k} \approx 2^n(n \log 2 + 0.577),$$

since it may be shown from well-known results that for any $n > 0$

$$n \log 2 + 0.577 \leq \sum_{k=1}^{2^n} \frac{1}{k}.$$

4 Parallel Hierarchical Learning

We turn now to a brief treatment of parallel hierarchical learning. The general assumption is that at each level of the hierarchy all functions at that level are being learned simultaneously, i.e., on the same trials.

To provide some immediate intuitive comparisons, consider the hierarchy of Figure 1 with 8 variables. As discussed earlier, $N_t = 256$ and $S_t = 28$, but $P_t = 12$, i.e., just 4 trials for each level of the hierarchy, of which there are 3. For that same structure generalized to n variables, as previously noted $N_t = 2^n$ and $S_t = 4(n-1)$. We state the result for P_t as a theorem.

Theorem 4.1 *With learner control of pattern selection at each level for the hierarchical pattern of Figure 1 generalized to n variables, $P_t = 4\log_2 n$.*

PROOF. The proof is quite simple. The four selected patterns for parallel processing at the initial stage and recursively then for each next level have the simple form of presenting the same values to every binary function at a given level. For example,

x_1x_2	x_3x_4	....	$x_{n-1}x_n$
11	11	......	11
10	10	......	10
01	01	......	01
00	00	......	00

Finally, it is easy to see that for this special hierarchy the number of levels is $\log_2 n$.

The reduction from $N_t = 2^n$ to $P_t = 4\log_2 n$ is dramatic, but of the sort required for massive parallel networks like the human visual system to be feasible. Of course, the strict condition of learner control of pattern selection at each level of a hierarchy will never be satisfied in nonexperimental natural environments where the distribution of stimulus patterns is not under learner control. Moreover, this natural distribution also fixes the sampling at every level of the hierarchy.

Even under desirable sampling distributions such as the uniform distribution, whether independent sampling occurs at each level or only once at the bottom level can, as would be expected, make a big difference in the rate of learning. Referring once again to the type of hierarchy shown in Figure 1 generalized to n variables, we make the following specific additional assumptions to illustrate the impact of this variation in sampling patterns.

(i) The 2^n stimulus patterns are divided into $2^{\frac{n}{2}}$ classes and each class contains exactly $2^{\frac{n}{2}}$ patterns.

(ii) For each class the functions $f_i, i = 1, \ldots, 2^{n-1}$ at the first level above the bottom are the same Boolean function for all patterns in a class.

(iii) But all 16 binary Boolean functions are used by the various classes, with a uniform distribution across classes.

For serial hierarchical learning with independent uniform sampling for each binary function at every level the expected number of trials for complete learning is given above, namely, $4(n-1)$, and for parallel processing under the same sampling assumption, $4\log_2 n$.

In contrast, when sampling occurs only at the lowest level of the hierarchy—a realistic assumption in many environments—, but is uniformly distributed at that level, and the hierarchy satisfies conditions (i)–(iii), then the expected number of trials for complete learning is bounded from below by

$$2^{\frac{n}{2}} \sum_{k=1}^{2^{\frac{n}{2}}} \frac{1}{k},$$

since a necessary condition of learning is that at least one stimulus pattern must be sampled from each class.

The nature of this example shows that even with parallel hierarchical processing, if the sampling pattern is determined only by sampling at the lowest level of the hierarchy, then in relatively simple cases learning will take exponentially many trials. The strongly contrasting results for independent sampling at each level of the hierarchy suggest why the teaching of any complex subject is broken up into independent stages of learning.

Stanford University, USA

References

[1] A. Blumer, A. Ehrenfeucht, D. Haussler and M. K. Warmuth. Learnability and the Vapnik–Chervonenkis dimension. *Journal of the Association for Computing Machinery*, **36**, 929–965, 1989.

[2] R. L. Rivest and R. Sloan. Learning complicated concepts reliably and usefully. [Extended Abstract]. *Proceedings of the 7th National Conference on Artificial Intelligence*, **2**, 635–640, 1988.

[3] L. G. Valiant. A theory of the learnable. *Journal of the Association for Computing Machinery*, **27**, 1134–1142, 1984.

[4] L. G. Valiant. Learning disjunctions of conjunctions. *Proceedings of the 9th International Conference on Artificial Intelligence*, **1**, 560–566, 1988.

5

Inference, Methodology and Semantics

Johan van Benthem

1 Breaking Up?

Of old, there has been an alliance between modern Logic and the Methodology or Philosophy of Science. Famous philosophers–logicians like Carnap, Beth or Reichenbach bridged the gap between these two disciplines: whose concerns, indeed, formed a continuum in their work. Nowadays, however, it seems fair to say that this particular contact has receded into the limelight, whereas new interfaces between Logic and Linguistics or Computer Science are making the headlines. (Significantly, the well-known *Handbook of Philosophical Logic* devotes not a single paper in its applied final volume to the above connection.)

In the eyes of many people, there is indeed a clean break here. Science is concerned with the deeper Reality behind the surface phenomena around us, while Linguistics investigates natural language, that is, merely the prima facie vehicle for expressing our pre-scientific 'common sense' views of the world. Moreover, Computer Science merely adds some further details of mechanical implementation to all this. Thus, there has been a (not so exceptional) switch of partners, and all that remains is to part in style.

The thesis of this paper is different, however: namely, that there remains a lot of intrinsic continuity between the earlier concerns of scientific methodology and those pursued under the newer banners.

This overlap is just a plain fact, which may be observed with many specific topics of research. A prime example is the study of *conditionals*, with its successive contributions by Goodman, Lewis or Adams: where issues of conditional reasoning, causality and lawlikeness all occur intertwined, so that it would be hard to say where Philosophy of Science ends and Logic or Linguistics begins (cf. the anthology of Sosa [29]). And significantly, this research line has only received further impetus in the present computational phase, witness such studies as Gärdenfors [14] or Makinson [22]. Another noteworthy example is the study of the logical structure of scientific *theories*, with classical contributions by Ramsey, Carnap, Suppes and Sneed. Here again, similar structures have come up in computational settings.

P. I. Bystrov and V. N. Sadovsky (eds.), Philosophical Logic and Logical Philosophy, 63–82.

In many respects, such theories are very much like databases, knowledge bases or abstract data types. This may be seen again in [14], where an analysis of theory cumulation and revision turns out to apply equally well to truth maintenance in data bases. Another example is van Benthem [11], which shows in some detail how the classical discussion of the relative status of 'observational' and 'theoretical' vocabulary in scientific theories reappears in the semantic literature on abstract data types, which also distinguishes between several layers of 'visible' versus 'hidden' predicates and functions.

But, this continuity can also be understood at a more general level. What was kept alive in this century by intellectual figures like the above three was a broad conception of Logic as a general methodology of reasoning, beyond the confines of the narrower 'mathematical turn' (no matter how successful in itself) occurring in the foundations of Mathematics. And this broader conception naturally includes a spectrum of cognitive activities, all the way from the realm of Common Sense to that of Science. Not surprisingly then, some of the well-known analyses of logical structures given by the logical positivists, even when doing less than justice to the full intricacy of scientific reasoning, are rather good models of general features of such ubiquitous phenomena (also in ordinary rational life!) as explanation, confirmation or refutation.

In what follows, we shall try to demonstrate this continuity in more detail, and attempt an assessment of the resulting challenge for methodologists and philosophers of Science. There is no pretence here at giving a complete survey, merely an attempt at increasing the weight of the evidence. This will be done under three headings, stressing, successively, general modes of inference, the ontological structures behind them, and their computational aspects.

The reason for choosing this particular subject in this particular book has to do with what I take to be a long-standing interest of Professor Smirnov's: namely, the bridging of the gap between contemporary Logic and general Methodology (cf. [28]). Moreover, there is also a personal reason, in that my own research has moved the same way, from a view of the Philosophy of Science as a 'most favoured trade partner' of Logic (cf.[3]) to the wider landscape sketched above (cf. [7]). And looking back after a number of years, this break appears less radical than it might have seemed at first glance. In many ways, one is pursuing the same ends, be it by different means.

2 Varieties of Inference

The classic achievement of modern Logic has been the formal explication of two notions of valid consequence: one axiomatic-deductive ($\vdash$), in the style of Frege and Hilbert, another semantic ($\models$) in the style of Tarski. And the question as to their mutual relation sparked off the first truly significant meta-logical result, namely Gödel's Completeness Theorem stating that the two declare the same inferences valid in such a widely applicable formalism as elementary predicate logic. Thus, we have a standard account of the notion of a valid inference step, in the format

From premises $P_1, \ldots, P_n$
to some conclusion C.

But in the practice of reasoning, this is only one building block in a variety of forms of rational argument. For a more complex mode of inference, let us take a look at a central theme in the Philosophy of Science.

Explanation as a Mode of Inference

When is a phenomenon B explained by certain facts or observations A? Perhaps the simplest answer is to present a deductive relation $A \vdash B$. But this is inadequate for various reasons. For a start, explanation takes place against the background of some general theory T licensing the transition from A to B:

$$T \wedge A \vdash B.$$

Moreover, in order to avoid triviality, A has to be really necessary to this derivation:

$$T \nvdash B.$$

And finally, the explanatory material should be at least consistent with the background theory:

$$T \nvdash \neg A.$$

There are intuitive problems still, even with this account of explanation: but at least, it may serve as a more realistic model of what goes on. It remains to add one further realistic touch. In many cases, the link between A and B may be less rigid than deductive consequence. Often, even in scientific explanation, some 'auxiliary hypotheses' have to be made, expressing a *ceteris paribus* rider to the effect that a number of unusual invalidating circumstances do not occur. A classical example is the event of an explosion, which may be explained, given the laws of Physics, by the earlier striking of a match in a room filled with gasoline vapours, under the auxiliary hypotheses that oxygen was present, the match was not defective, etc. Thus explanations may occur in various degrees of deductive strength.

This single example already demonstrates a variety of logical issues. For instance, what emerges here is a richer picture of a logical inference than the bare schema

$$\frac{P_1, \ldots, P_n}{C}$$

In fact, one is rather reminded of Toulmin [33] in itself a highly polemical book trying to undermine classical logic, by proposing the following central pattern of human reasoning:

		Qualifier	
Data	———————→ Warrant		Claim
		Rebuttal	

But, this purported competitor to the earlier standard schema of inference is in fact almost item-by-item the above schema of explanation, witness the following transcription

		'Deductive Strength'	
Explanans	———————→ Background Theory		Explanandum
		'Auxiliary Hypotheses Refuted'	

Moreover, some further important technical matters are raised by the above. Two notable examples are the choice of Deductive versus Default reasoning for driving explanation, as well as that of Relevance of explanatory premises to their conclusions.

For the moment, however, we highlight another issue, namely, the more traditional 'logic' of the notion of explanation formulated above: Fixing a background theory T, one may consider the transition from A to B in the light of the usual properties of logical valid consequence. And them we find certain differences with the less elaborate standard pattern. For instance, on the one hand, there is still *Transitivity* ('transmission of explanation'):

If A T-explains B and B T-explains C,
then A T-explains C.

But on the other hand, other basic properties, such as *Monotonicity* in either premises or conclusions is lost. It may easily be checked that

If A T-explains B,
then it neither follows that $A \wedge C$ T-explains B, nor that A T-explains $B \vee C$.

On the positive side again, what explanation retains are two other well- known features of classical consequence, being Conjunction of conclusions and Disjunction of premises.

So we see that various modes of inference may show different formal properties. And the latter should be studied systematically, in the spirit of Bernard Bolzano's historical approach to Logic (cf. [6] which also point out how Bolzano's logical program fits in with his theory of science). In other words, what we need is a truly

General Theory of Inference

What will be considered now is a broad investigation of consequence relations, together with their logical properties.

A new impetus for this kind of research has come from recent developments in Artificial Intelligence, where a much greater variety of modes of inference than just standard consequence has turned up in the design of expert systems or intelligent planning systems. One systematic program in this spirit may be found in [22], which classifies various kinds of non-standard inference by logical properties in the above vein. (See also [35] for a related, more traditional Polish approach.)

Here again, History turns out to have a role to play. There is a useful analogy between the above concerns and an earlier research line in Philosophical Logic, namely the study of counterfactuals and conditionals ($\Rightarrow$) in general. As is explained in [11], many interesting properties of non-standard inference are supplied by the tradition of *Conditional Logic*. Notably, the latter modes of inference do not just lose familiar features from standard systems (as is over-dramatised in the current negative appellation of 'non-monotonic logic'). For already the so-called 'minimal conditional logic' discovered in the tradition of Lewis, Stalnaker and Pollock, offers various substitutes for classical properties lost. An example is the following somewhat more subtle form of leftward Monotonicity, which also turns out to hold for he above notion of explanation:

$$\text{If } A \Rightarrow B \text{ then } A \wedge (B \vee C) \Rightarrow B.$$

What this says intuitively is that premises may still be strengthened with any statement which is a standard consequence of the conclusion.

What we shall do here, however, is to pursue, not this proof-theoretic track, but rather a model-theoretic format suggested by the semantics of natural language: in particular, the study of so-called 'generalized quantifiers' (cf. [7]).

Before going into technical details, let us consider some general analogies here. In the semantics of natural language, one studies categories of expression, both by themselves, and in their interaction. Now the first step in the semantic analysis of any given grammatical category is the location of some suitable semantic scheme for its interpretation. For instance, with the central case of determiner expressions, such as 'every', 'three', 'most', or 'no', this turns out to be that of binary relations between unary predicates. Then, within this scheme, specific truth conditions may be stated for individual lexical items of the relevant category. But, perhaps the more interesting semantic task is to find broad semantic characteristics of certain natural *families* of expression, such as 'quantifiers' or 'locatives' or 'absolute adjectives'. These come in the form of so-called *denotational constraints*, governing the admissible meanings for expressions in the family. For instance, all linguistic determiners seem to obey the requirement of 'Conservativity', expressing a 'domain restriction' to their left argument:

$$\text{Det } AB \quad \text{iff} \quad \text{Det } A(B \cap A).$$

And the more narrowly logical determiners among these satisfy an additional requirement of 'invariance for permutations' π of the universe of discourse:

$$QAB \quad \text{iff} \quad Q\pi[A]\pi[B].$$

This makes them purely 'quantitative', depending only on numbers of individuals inside their argument predicates.

Now, the semantic study of inference, we submit, lies at a similar 'intermediate' level of analysis: calling for an interplay between specific explications for particular candidates, and broad semantic constraints on 'the family' as a whole. Further concrete examples of this perspective will be presented now, exploiting what is at least a formal analogy with the case of determiners.

A Semantic Perspective

A general semantic setting for logical speculation about notions of consequence is provided by *Venn Diagrams* of the following form

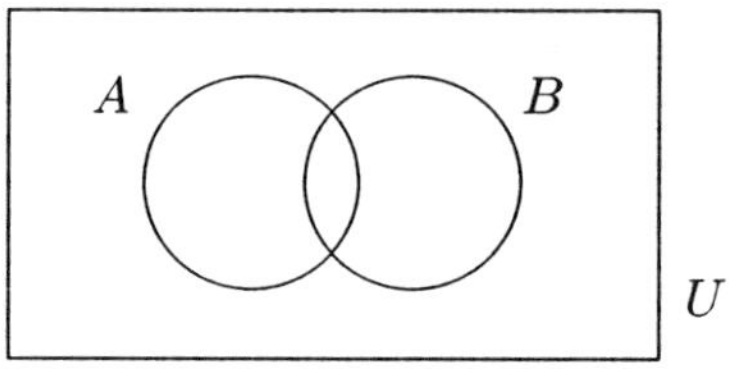

Here U is the total universe of relevant models ('worlds', 'situations', 'states'), while A represents those models where the antecedent holds, and B those where the consequent holds.

Now, one can experiment with intuitions of 'conditionality' by imposing various semantic constraints on the admissible positions of A and B, in order to have a conditional connection between the two. For instance, there may be upward monotonicity in the consequent, in that enlarging the area of B will not disturb the truth of the conditional assertion. And more generally, any change in the picture which favours the 'examples' $A \cap B$ over the 'counter-examples' $A - B$ will be allowed. See [5] for details, as well as some technical results, classifying all possible conditionals within a number of semantic constraints. (Incidentally, the analysis in the latter paper draws upon observations coming both from Linguistics and the Philosophy of Science: more precisely, Carnap's analysis of the role of *dispositional statements* in scientific theories.)

Especially since there need not be one single definition of conditionality or consequence capturing the essence of this notion once and for all, the present format seems to provide a suitable level of generality in our philosophising. For instance, one may try to formulate general semantic constraints like the above, seeing which precise candidates are still admitted by them. But also, one can start from specific candidates—for instance, the set-theoretic *inclusion* $A \subseteq B$ corresponding to classical Tarskian entailment—and try to distinguish between its general traits and its private peculiarities.

• Let us illustrate this line of thinking first for another well-known notion from methodology, namely that of *Confirmation* of a hypothesis by certain evidence.

As with Explanation, there is a background theory here, guiding the transition from the latter to the former. Thus, the relevant model- theoretic notion is a *ternary* relation between classes of models, whose Venn diagrams look as follows:

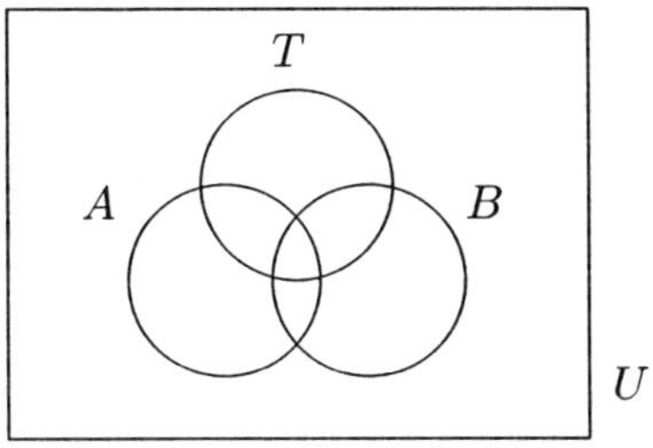

Now, this ternary relation is *context-independent*, in that it does not really depend on the surrounding universe U. Borrowing a notion from logical semantics, we may state this as a principle of *Context Neutrality*:

$$\text{Whenever } U \subseteq U', \text{ then } \mathbf{conf}_U \mathrm{TAB} \text{ iff } \mathbf{conf}_{U'} \mathrm{TAB}.$$

Moreover, instead of U , it is rather the background theory T which sets the stage here. This may be formulated in analogy with the already mentioned constraint of *Conservativity* governing natural language determiners, being an instance of the general linguistic phenomenon of 'domain restriction':

$$\mathbf{conf}_U \mathrm{TAB} \text{ iff } \mathbf{conf}_U T(T \cap A)(T \cap B).$$

Together, the two constraints so far allow us to re-present Confirmation as follows:

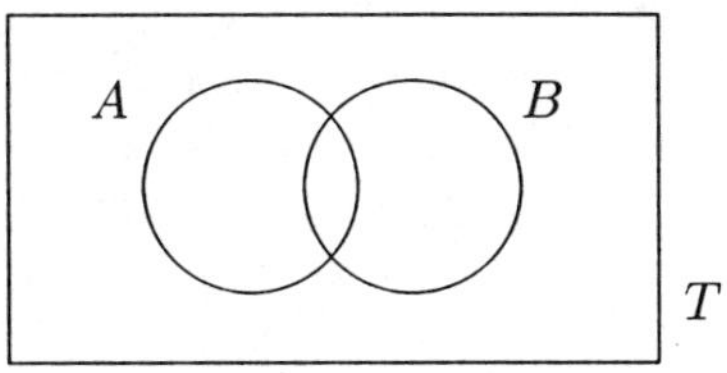

Next, Confirmation intuitively satisfies further semantic constraints which make it very similar to the earlier notion of 'conditionality'. In particular, it seems reasonable to demand *Monotonicity* in the hypothesis verified:

$$\text{If } \mathbf{conf}_T AB, \text{ then } \mathbf{conf}_T A(B \cup C).$$

Moreover, decreasing the zone of counter-examples will never do any harm — and so we have *Cautious Persistence:*

$$\text{If } \mathbf{conf}_T AB \text{ and } X \supseteq B, \text{ then } \mathbf{conf}_T(A \cap X)B.$$

Finally, there is the question as to which combinations of different cases of Confirmation are admissible. And here, a case can be made for the following two:

Conjunction of Hypotheses

$$\text{If } \mathbf{conf}_T AB \text{ and } \mathbf{conf}_T AC, \text{ then } \mathbf{conf}_T A(B \cap C),$$

Disjunction of Evidence

$$\text{If } \mathbf{conf}_T AB \text{ and } \mathbf{conf}_T BC, \text{ then } \mathbf{conf}_T(A \cup B)C.$$

Within these constraints, the range of possible specific definitions of Confirmation may be investigated using the techniques of van Benthem [5, 7]. In particular, the case of 'bare sets' of models, where permutations of the individual models do not affect the confirmation relation between A and B allows essentially just *inclusion*, i.e. classical entailment.

More interesting notions only arise when some additional structure on these models is postulated, say, a binary relation of *preference*. In that case, one reasonable option is to require that all 'most preferred' A- models be B-models — and all the above formal principles would still hold. Significantly, the latter strategy reflects precisely the well-known possible worlds semantics for Conditional Logic, as developed by Lewis and Stalnaker. ([11] also establishes a connection with the recent computational theory of Circumscription, tracing its ancestry to a semantics of Confirmation proposed already in [16].)

• Now, let us return to the notion of *Explanation*. Here again, the earlier constraints of Context Neutrality and Conservativity seem to apply, whence there is a reduction again to a ternary relation $\mathbf{exp}_T AB$. But then, keeping the earlier specific proposal in mind, there seems no intuitive evidence for requiring Monotonicity in any of the three arguments. More plausible, perhaps, are the two combination principles of Conjunction and Disjunction. For instance, if some observation explains both B and C, does it not also explain the conjunction $B \wedge C$ then? To get a better feel for the possibilities here, one may also look at the specific proposal made earlier, which amounts to the following requirement on the Venn Diagram:

$$T \cap A \subseteq B \qquad T \cap A \neq \varnothing \qquad T - B \neq \varnothing.$$

It would be of interest to find a complete axiomatization of the inferential principles governing this notion.

One way of looking at the situation is this. For classical entailment, represented by the first semantic condition only, we know what the characteristic properties are: namely, Reflexivity, Transitivity, left- and right-Monotonicity (as well as Conjunction of conclusions and Disjunction of premises). Adding the second condition takes us to Bolzano's notion of valid logical inference, for which [6] records a loss of Reflexivity and left Monotonicity, which are to be replaced by such more cautious variants as

$$\frac{A \Rightarrow B}{A \Rightarrow A} \qquad \frac{A \Rightarrow B}{B \Rightarrow B} \qquad \frac{A \Rightarrow B \qquad A \wedge C \Rightarrow A \wedge C}{A \wedge C \Rightarrow B}.$$

What the third semantic requirement then does is to leave these new variants in force, while replacing right Monotonicity by a similar cautions rule:

$$\frac{A \Rightarrow B \qquad B \vee C \Rightarrow B \vee C}{A \Rightarrow B \vee C}.$$

Finally, again, it may be reasonable eventually to pass to a more structured universe of models, carrying a binary relation of preference. For instance, then, explanations involving less than one hundred percent deductive certainty may be modelled by requiring only that the most preferred A- models are B-models inside the T-domain.

• As a final example of this type of analysis, we mention the notion of *Verisimilitude* between theories T and two of their competing approximations A, B. Here, a well-known specific explication is the Miller–Kuipers notion of 'A is at least as good an approximation to T as B is':

$$T \cap B \subseteq T \cap A \quad \text{and} \quad A - T \subseteq B - T$$

The relevant family of notions is analysed in the above setting in [8]. In this case, the semantic constraints are quite different from the above, witness conditions like the following:

$$\begin{array}{lll} \mathbf{ver}_T AB & \text{iff} & \mathbf{ver}_T(A - B)(B - A) \\ \mathbf{ver}_T AB & \text{iff} & \mathbf{ver}_B AT. \end{array}$$

But, the general spirit of the investigation remains similar.

Conclusion

The outcome of our discussion so far has been this. Methodology is concerned with a variety of cognitive procedures, that are inferential in a broad sense. This calls for a systematic logical study of various natural options, and their theoretical properties. And suitable perspectives here will arise only in a liberal co-operation between Philosophy of Science, Philosophical Logic and Logical Semantics.

Eventually, various other mechanisms should be included in this enquiry too. A case in point are again the earlier conditionals. One plausible view of so-called 'counterfactual conditionals' is that these involve inference from premises which have first been *revised* so as to admit the antecedent assumption consistently. But then, the process of revision of premises or information states should be analysed in equally general terms. (See [14] for an elegant formal theory.) Such revisions also come into play when we take the currently emerging 'dynamic' point of view that certain types of information, such as generic statements or other default rules, amount to instructions for changing our expectations or preference patterns over universes of models. (See [30, 34] for formal semantic theories of the latter kind.)

3 Varieties of Ontology

Contacts between logical semantics and the methodology of science are not confined to notions of inference or more general cognitive procedures. They also extend to the epistemic structures on which these procedures are supposed to operate as well as their underlying ontology.

A case in point is the study of *temporal* structures in natural languages and computation. In order to make statements about events taking place over time, and drawing inferences about them, a rather diverse ontological picture is needed of various temporal representations and their connections, witness [18]. (See [4, 9] for a survey.) And this leads us directly to issues in the Philosophy of Science. For, a proper account of temporal meanings in natural language seems to involve both more scientific 'point based' and more common sense oriented 'period based' paradigms. This duality between various temporal ontologies has also been prominent in Computer Science, witness [1] or [21], which are concerned with the interplay between actual physical time and the representation time of our computational models.

This is one instance of a more general theme. In fact a discipline like Measurement Theory (see [19]) may be viewed as being concerned precisely with the bridge between the more qualitative representation structures underlying the common sense world of natural language and the quantitative structures of science. This general perspective is still far from complete, and the ontological map retains its uncharted territories. For instance, there is another intriguing broad duality in natural languages, whose systematic relation to general reasoning and scientific theory formation has not been investigated yet. This is the distinction between so-called *count terms* and *mass terms* in describing the world (cf. the survey chapter by [26]). The former refer to concrete countable representations of the world into 'guns', 'days' or 'glasses', whereas the latter refer to continuous quantities like 'wine', 'time' or 'patience'. Both have their own forms of quantification: witness 'much time' versus 'many guns', or 'ten days' versus 'ten glasses of wine'. The existence of such parallel systems in the common sense world seems significant, even though we still do not understand its precise cognitive purpose.

For another kind of example, we return to the earlier-mentioned quantifier system of natural languages, showing how apparently purely semantic questions can lead us back to the foundations of Science.

Quantification and Probability

One common non-standard determiner expression is 'almost all A B', which tells us that only very few exceptions may lie in the Venn zone $A - B$. The intuitive logic of this relation Q satisfies the following evident properties, for sufficiently large domains A:

1. QAB implies $QA(B \cup C)$ — Right Monotonicity
2. QAB, QAC imply $QA(B \cap C)$ — Conjunction
3. $QA(A - \{x\})$ — Non-Atomicity
4. *not* $QA\varnothing$ — Non-Triviality

Moreover, recall the earlier requirement on logical determiners of Permutation Invariance, making them sensitive only to numbers of individuals in Venn zones. What this implies already is that the left domain A must be *infinite*, by the last three properties. For instance, one candidate which satisfies all requirements is the quantifier 'all, with at most finitely many exceptions'. Finally, once one considers multiple quantifier combinations, involving also binary or higher predicates, there is another natural condition, expressing that 'almost all' has the same 'scope freedom' as the universal quantifier. In an ad-hoc notation:

5. $Q^A x \cdot Q^A y \cdot Rxy \quad \text{iff} \quad Q^A y \cdot Q^A x \cdot Rxy.$

Now, the resulting system of axioms has in fact been proposed by Harvey Friedman as a rendering of the main properties of the *probabilistic quantifier* 'with at most a measure 0 set of exceptions'. (Cf. [20].) In particular, the final axiom proposed above is then a form of the well-known Fubini Theorem in Probability Theory. And indeed, we may show how this total set of conditions demands a certain probabilistic structure on the universe of individuals, by forcing us to leave the realm of purely numerical logical determiners in the above sense. (This supplies a proof announced in [10].)

Proposition 3.1 *No logical generalized quantifier satisfies the Friedman Axioms for all set arguments.*

PROOF Let us consider even the special case where the left-hand argument is the whole universe. Thus Q becomes a unary quantifier, being a family of sets. It suffices to prove the following

Claim 3.2 (AC). Such a quantifier is closed under the formation of subsets.

For, then, if $A \in Q$ (this happens for some A, by Axiom 3), then $\varnothing \in Q$, which contradicts Axiom 4.

PROOF OF CLAIM By induction on the cardinality of B (the Axiom of Choice is presupposed) we show that the following equivalent formulation holds:

$$A \in Q, B \subseteq A \quad \text{imply that} \quad (A - B) \in Q.$$

Case 1. *B is finite.*

Finite parts may be subtracted using Axiom 3 plus Axiom 2.

Case 2. *B is infinite with $|B| = \mathbf{m}$.*

The inductive hypothesis reads like this:

$$\forall A \in Q \ \forall B \subseteq A : |B| < \mathbf{m} \Rightarrow (A - B) \in Q.$$

Now, well-order A as follows

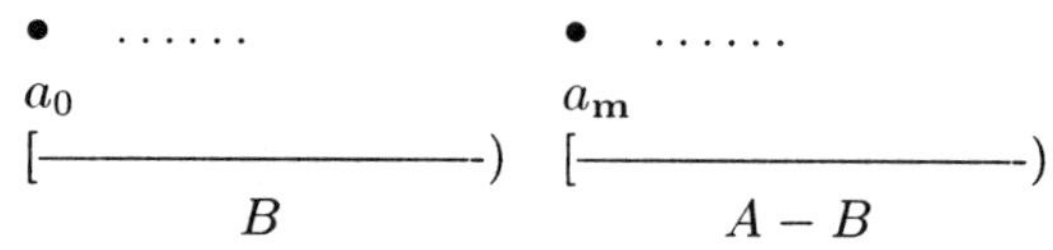

And, define a binary relation R on A via

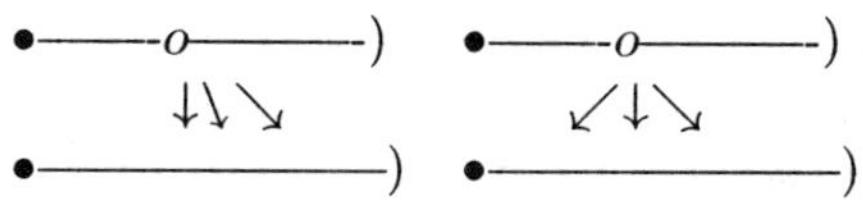

where

$$Ra_\alpha y \quad \text{iff} \quad \begin{array}{l} (1)\ \alpha \geq \mathbf{m}\ \&\ y \in A, \text{or} \\ (2)\ \alpha < \mathbf{m}\ \&\ y = a_\beta \text{ with } \beta \geq \alpha. \end{array}$$

Then we have

$$\forall x \in A\ Qy \cdot Rxy.$$

To see this, observe that

If $x = a_\alpha, \alpha \geq \mathbf{m}$, then $(R)_x = A$: which is in Q.
If $x = a_\alpha, \alpha < \mathbf{m}$, then $(R)_x = A - \{a_0, \ldots, a_\alpha\}$:
and since the set subtracted here has cardinality $< \mathbf{m}$, $(R)_x \in Q$
because A is, by the inductive hypothesis.

Therefore also, because $A \in Q$, Axiom 1 implies that

$$Qx \cdot Qy \cdot Rxy$$

Now apply Axiom 5: we must also have

$$Qy \cdot Qx \cdot Rxy.$$

Or spelt out: $\{y \mid \{x \mid Rxy\} \in Q\} \in Q$. So, which individuals y have ${}_y(R) \in Q$?

If $y = a_\alpha$ with $\alpha < \mathbf{m}$, then ${}_y(R) = (A - B) \cup \{a_\beta \mid \beta < \alpha\}$.
If $y = a_\alpha$ with $\alpha \geq \mathbf{m}$, then ${}_y(R) = A$: which is in Q.

Now, once again:

$$\{y \mid_y (R) \in Q\} \in Q.$$

Case 1. This set equals $A - B$, and we are done: $(A - B) \in Q$.

Case 2. This set exceeds $A - B$, and there exists at least one $y \in B$ such that $_y(R) \in Q$. By the above observation, $_y(R) = (A - B) \cup$ 'some set of cardinality $< \mathbf{m}$'. By the inductive hypothesis then, applied to $_y(R)$: $(A - B) \in Q$, and we are done again.

Thus, a semantic analysis of common patterns of quantification leads us naturally to investigate such methodological issues as the logical genesis of probabilistic reasoning. In addition to such examples coming from special linguistic categories, one can also look at the categorial structure of natural languages as a whole. For instance, [24] shows the penetration of linguistically inspired 'categorial languages' into the analysis of scientific theories.

Conclusion

Once again, some general morals may be drawn here. The semantic analysis of specific linguistic categories of expression naturally merges with known methodological issues, witness the cases of conditionals, temporal expressions and quantifiers. And more generally, a comparison between the two research perspectives seems useful too. In particular, there is a common systematic task, namely the exploration of a space of plausible ontologies and their connections, in order to provide a flexible instrument for representing Reality at various levels of 'grain size'.

4 Computation

The emphasis so far has been placed on cognitive procedures and the structures involved in these. But then, it is only a small step towards actual concerns of computation and suitable data structures. Therefore, it comes as no surprise that several theories developed in the methodology of Science also admit of a more computational interpretation.

As for data structures, we already mentioned the analogy between scientific theories and data bases, or more generally, knowledge states, as developed in [14]. For instance, well-known accounts of the structuring of scientific theories into more 'negotiable' and more 'entrenched' layers have their counterparts in recent theories about levels of entrenchment in knowledge bases. And also the traditional distinction between theoretical versus observational terms in the vocabulary of scientific theories found precise analogies in the theory of abstract data types (cf. [11]). Finally, another illustration is the use of Carnap's work on theory structures in the computational setting of [12].

As for the more procedural side, a computational concern with algorithms operating on knowledge structures has a natural basis in general methodology too. A rather down-to-earth illustration is the use of rules from Carnap's Inductive Logic in a well-known expert system like MYCIN. Deeper connections may be found too, for instance in the recent influential treatment of knowledge and causality for planning systems in [27].

There is also an intrinsic basis for such computational concerns within Methodology and Logic itself. For instance, an 'explanation' in the usual sense of the term is not so much a statement as an actual *argument*. And one step further in this realistic direction, there is of course a systematic ambiguity in our very use of such terms as 'explanation', 'judgement' or 'deduction': as these may denote both the *activity* and its *products*. And there is much to be said for retaining this intuitive association. (Again, [33] already had a significant thesis here: namely, that logic should be concerned with procedural 'formalities' as much as with linguistic 'form'.) This tendency can trace its roots back to long-standing constructive views of Mathematics, where proofs and calculations are intertwined. But it also corresponds to evident features of Natural Science, where measurement, testing and experimentation is of the essence. For earlier logical paradigms motivated along these lines, see [23], but also e.g. the general 'operational' perspectives on physical theories presented in the collection [17]. (Also congenial is the operational Skolem function approach to scientific theories advocated in [32].) To put the issue more pregnantly, where would a non-computable 'solution' to a problem be acceptable at all in Science (except for pure Mathematics)?

Complexity and Fine-Structure

What all this involves at least is attention for the *complexity* of the earlier notions, as an intrinsic feature of their logical status, and for the development of recognising algorithms in general.

Remark 4.1 (Complexity of Methodological Notions). Actually, the earlier notions suffer from defects in this respect. For instance,

- 'Explanation' for formulas of predicate logic is an undecidable, and even non-axiomatizable notion.

The reason lies in the following reduction. Setting T = TRUE, $\mathbf{exp}_T AA$ will hold if and only if A is a *contingent* formula of predicate logic. And the latter class of formulas is not recursively enumerable.

Another illustration may be taken from the theory of Definability, which is so prominent, e.g. in [28]. Despite the central importance of Beth's Definability Theorem, it also has this darker side that

- For finitely axiomatized first-order theories $T(\mathbb{P}, Q)$ and predicates Q, the question whether Q is explicitly $\mathbb{P}$-definable in T, though recursively enumerable, is *undecidable*.

This may be seen by reducing universal validity in predicate logic (itself an undecidable notion) effectively to a definability question of this kind:

PROOF. Let A be a formula in the vocabulary $\mathbb{P}$. Choose some new proposition letter (0-ary predicate) q not occurring in A. Then, the following two assertions are equivalent:

1. A is universally valid.

2. q is explicitly $\mathbb{P}$-definable in the theory $\{A \to q\}$.

From 1 to 2. In this case, TRUE is an explicit definition for q in $\{A \to q\}$.

From 2 to 1. Suppose that $\{A \to q\} \models q \leftrightarrow B$ for some $\mathbb{P}$-formula B not containing q. Then (i) $q \models q \to B$ — i.e. $q \models B$ — and hence $\models B$ (since q does not occur in B). Next, also (ii) $\neg A \models q \leftrightarrow B$. Combining (i) and (ii) then, we have that $\neg A \models q$. And, since $\neg A$ does not contain q either, it follows that $\models A$. ∎

These 'negative' outcomes do not means that computational concerns had better be left out of our methodology: that would be sticking one's head in the sand. And more positively, despite this complexity, a general computational perspective also leads to genuine enrichment. For instance, one natural and useful theme will become that of *fine-structure*. In computation, it is essential to consider finer details of the syntactic and semantic architecture of general purpose formalisms, asking for their various layers of expressive and computational power. An example is the restriction to such fragments of predicate logic as the *Horn Clause* formalism in Logic Programming, whose algorithmic behaviour is much more manageable in practice. Interestingly, a similar distinction may be observed in traditional accounts of scientific theories, where the 'theoretical laws' might have any predicate-logical complexity (though reducible, in principle, to universal Skolem forms) whereas the 'empirical generalizations' often look like universal Horn clauses, admitting of operational procedures. Another illustration would be the above-mentioned theory of Definability, where one might rather be interested in definitions falling inside certain simple natural fragments of predicate logic (say, the finite variable fragments corresponding to intensional formalisms). In particular, there would be just as much interest in *negative* as in positive results, for instance, when locating intrinsic jumps in expressive power needed to explicate some methodological notion.

Models versus Proofs: the Case of Relevance

To conclude here, we consider another thread from our earlier discussion. The 'standard history' of the Philosophy of Science, as taught in many universities, is one continuous progress toward Enlightenment. First there were philosophers concentrating exclusively on syntax and axiomatic deduction (such as Carnap or Hempel). Then, a model-theoretic perspective dawned, bringing in semantic structures (as heralded by Beth). Afterwards, it was realised by Suppes and Sneed that one can throw away the dead weight of the linguistic formalism, concentrating directly on set-theoretic structures, the way ordinary scientists do. But with the advent of computational concerns, the wheel of History makes another revolution: and we need *code* on which to perform our calculations. Thus, the classical 'syntactic' phase gets rehabilitated.

Of course, there is also a new twist to the classical discussion. In particular, there will now be an emphasis on syntax plus *proof* structure, as was already observed above. For instance, it seemed that various crucial features of Explanation or Confirmation do not yet come out correctly on the earlier semantic analysis. Notably, there is the issue of Relevance of evidence to hypotheses, or of explanantia to explananda. This rather seems to involve a reference to the actual *argumentation* brought forward in order to establish their connection.

There is a very general issue emerging here. The usual tradition has been to analyse explanation and similar modes of inferential connection as a relation between *statements*. But perhaps, the more natural use of this notion applies to *arguments*. What explains the occurrence of the explosion in Section 2 is not the mere statement of sufficient preconditions for that event, but rather the whole argument from these conditions to the phenomenon to be explained. Put differently, the usual account is an instance of a rather common procedure in Logic, which may be called *The* $\exists$ *Strategy*:

'Suppress some loose parameter by an existential quantification'.

For instance, we say that B follows from A is there exists a proof, or that A is past if there exists a past moment where it occurred — whereas in practice, we will usually have some specific derivation, or specific reference time, in mind. These impoverished notions, where we 'hide' more specific information available, have their undeniable uses. But one should always keep in mind that there is more information available for logical analysis. (This point is made in [2].)

For instance, with the phenomenon of *Relevance*, which has played an important role in the methodological literature (cf. [15]), there is a strong case for taking a closer look at the actual arguments involved. So far, in Section 2, we have operated in a global manner, looking at ordinary consequences from premises to conclusions, and then 'filtering out' undesired cases. For instance, this was the point behind such additional stipulations as the consistency of the premises and the contingency of the conclusion. Or, already somewhat closer to the actual syntactic presentation of the premises, one obvious global filter would be this:

$$\mathbb{P} \models C, \qquad \text{but no proper subset of } \mathbb{P} \text{ implies } C \qquad *$$

But, such filters can be computationally costly. (For instance, [6] shows that the latter check makes relevant consequence non-effectively axiomatizable.) But then, it is also possible, and perhaps intuitively even more plausible, to impose effective conditions of relevance at the level of actual argument. Some *specific* derivation of B has been presented from $A_1, \ldots, A_n$, and the issue is whether it has used all of the premises mentioned in an essential manner. There are some obvious questions here for methodological reflection.

Digression 4.2 (Global and Local Relevance). The two perspectives mentioned just now are not obviously equivalent. This may be seen by comparing the above semantic notion $*$ with derivation in a natural deduction calculus, where book-keeping of actually used assumptions is straightforward. On the one hand, a derivation may use all of a certain set of premises, and that in a non-redundant manner (being in Gentzen 'normal form'), without establishing a $*$ relation, witness

$$\dfrac{\dfrac{A \quad (A \to (C \to B))}{(C \to B)} \qquad \dfrac{(A \to B) \quad ((A \to B) \to C)}{C}}{B}$$

Here, two of the premises would already suffice. On the other hand, not every $*$ relation admits of an irredundant natural deduction, witness the transition

$$A \models (B \to A)$$

which can only have a derivation involving a 'vacuous conditionalization'.

Much more can be said here, of course — including a discussion of the desirability of introducing some form of Relevant Logic here. But for the moment, it will suffice to have illustrated the issues involved.

Conclusion

Our conclusion must be that there is a natural transition from methodological to computational concerns. Or even more radically, that computational concerns form an essential ingredient of methodological investigations.

5 Staying Together?

The above has not been a complete survey of existing contacts between the Philosophy of Science and logical Semantics of Informatics. For instance, we have not gone into game-theoretical approaches to physical theories such as those of Stachow and Mittelstaedt in [17]. Nor have we considered such more algorithmic turns as the Learning Theory of [25] which has thrown an entirely new light on methodological discussions of the formation and modification of hypotheses and theories. of these contacts. It seems undeniable that linguistic and computational applications of Logic have engendered more sophisticated views of the logical apparatus itself, yielding many new insights into its fine-structure of expressive power and computational complexity. Probably, the logical study of scientific theories would benefit from the same sensitivity, reducing the danger of mere fomalization (which has lurked, for instance, behind broad frameworks such as Carnap's or Sneed's). Moreover, all this may surely be accomplished without loss of indigenous values.

And, in the opposite direction, one would hope that methodologists will join ongoing debates on the interface of language, logic and computation, such as that concerning the choice between Classical Physics and various forms of Common Sense Physics. Their expertise will be needed, just as much as that of pure logicians.

University of Amsterdam, The Netherlands

References

[1] J. Allen and P. Hayes. A common-sense theory of time. *Proceedings IJCAI 1985*, pp. 528–531, 1985.

[2] J. Barwise and J. Perry. *Situations and Attitudes*, The MIT Press, Cambridge, MA, 1983.

[3] J. van Benthem. The logical study of science. *Synthese*, **51**, 431–472, 1982.

[4] J. van Benthem. *The Logic of Time*, Reidel, Dordrecht, 1983.

[5] J. van Benthem. Foundations of conditional logic. *Journal of Philosophical Logic*, **13**, 303–349, 1984.

[6] J. van Benthem. The variety of consequence, according to Bolzano. *Studia Logica*, **44**, 389–403, 1985.

[7] J. van Benthem. *Essays in Logical Semantics*, Reidel, Dordrecht, 1986.

[8] J. van Benthem. Verisimilitude and conditionals. In *What is Closer-to-the-Truth?*, Th. Kuipers, ed, pp, 103–128, Rodophi, Amsterdam, 1987.

[9] J. van Benthem. Logic, time and computation. In *Linear Time, Branching Time and Partial Order in Logics and Models for Concurrency*, pp. 1–49, J.-W. de Bakker *et al.* (eds), Springer, Berlin, 1989.

[10] J. van Benthem. Polyadic quantifiers. *Linguistics and Philosophy*, **12**, 437–464, 1989.

[11] J. van Benthem. Semantic parallels. In *Logic Colloquium, Granada 1987*, pp. 331–375, H.-D. Ebbinghaus *et al.*, eds, North Holland, Amsterdam, 1989.

[12] J. Doyle. Some theories of reasoned assumptions: an essay in rational psychology. Department of Computer Science, Carnegie-Mellon University, Pittsburgh, PA, 1983.

[13] D. Gabbay and F. Guenthner, eds. *Handbook of Philosophical Logic, Vol. IV (Topics in the Philosophy of Language)*, Reidel, Dordrecht, 1989.

[14] P. Gärdenfors. *Knowledge in Flux. Modelling the Dynamics of Epistemic States*, Bradford Books/MIT Press, Cambridge, MA, 1988.

[15] C. Glymour. *Theory and Evidence*. Princeton University Press, Princeton, 1980.

[16] C. Hempel. *Aspects of Scientific Explanation*, The Free Press, Glencoe, IL, 1965.

[17] C. Hooker, ed. *Physical Theory as Logico-Operational Structure*, Reidel, Dordrecht, 1979.

[18] H. Kamp. Intervals, events and temporal discourse. In *Semantics from Different Points of View*, R. Bäuerle *et al.*, eds, pp. 376–417, Springer, Berlin, 1979.

[19] D. Krantz, D. Luce, P. Suppes and A. Tversky. *Foundations of Measurement*, Academic Press, New York, 1985.

[20] M. van Lambalgen. The axiomatization of randomness. *Journal of Symbolic Logic*, **55**, 1143–1167, 1990.

[21] L. Lamport. Inter-processor communication. Final report. SRI International, Menlo Park, 1985.

[22] D. Makinson. General non-monotonic logic. In *Handbook of Logic in Artificial Intelligence and Logic Programming*, D. Gabbay, C. Hogger and J. Robinson, eds, Oxford University Press, Oxford, 1989.

[23] Y. Medvedev. Finitnie zadachi. *Dokladi Akademia Nauk USSR*, **142**, 1015–1018, 1962.

[24] B. Mundy. Elementary categorial logic, predicates of variable degree and theory of quantity. *Journal of Philosophical Logic*, **18**, 115–140, 1989.

[25] D. Osherson, M. Stob and S. Weinstein. *Systems that Learn*, The MIT Press, Cambridge, MA, 1986.

[26] J. Pelletier and L. Schubert. Mass Terms. In *Handbook of Philosophical Logic*, vol. IV, D. M. Gabbay and F. Guenthner, eds. Kluwer Academic Publishers, Dordrecht, 1989..

[27] Y. Shoham. *Reasoning about Change: Time and Causation from the Standpoint of Artificial Intelligence*, The MIT Press, Cambridge, MA, 1988.

[28] V. Smirnov. *Logicheskije Metodi Analiza Nauchnowo Zanija*, Publishing House 'Nauka', Moscow, 1987.

[29] E. Sosa, ed. *Causation and Conditionals*, Oxford University Press, Oxford, 1975.

[30] W. Spohn. Ordinal conditional functions: a dynamic theory of epistemic states. In *Causation in Decision, Belief Change and Statistics, Vol III*, W. Harper and B. Skyrms, eds. pp. 105–134, Reidel, Dordrecht, 1988.

[31] F. Suppe, ed. *The Structure of Scientific Theories*, The University of Illinois Press, Urbana, IL, 1977.

[32] P. Suppes. Some open problems in the philosophy of space and time. In *Space, Time and Geometry*, P. Suppes, ed, pp. 383–401, Reidel, Dordrecht, 1973.

[33] S. Toulmin. *The Uses of Argument*, Cambridge University Press, Cambridge, 1958.

[34] F. Veltman. Update semantics for defaults. Institute for Language, Logic and Information, University of Amsterdam, 1989.

[35] R. Wojcicki. Aksiomaticheskaja traktowka nemonotonnich rassyzjdenij. In *Neklassicheskije Logiki i ich Primenenija*, V. Smirnov *et al.*, eds, pp. 140–144, Institute of Philosophy, Academy of Sciences of the USSR, Moscow, 1989.

6

Theories in Science

Ryszard Wójcicki

1 Theories vs. Deductive systems

From an informal point of view, an empirical theory is a set of empirical laws that govern phenomena of a certain specific kind. Thus, for instance, Mendelian Genetics is a theory of heredity, while Newtonian Particle Mechanics is a theory of mechanical movement of physical bodies.

There are two formal counterparts of this notion. Most often a theory is defined to be a *deductive system*, i.e. a deductively closed set of sentences of a certain fixed formalized language L. Alternatively, a theory may be defined to be a class of set-theoretic structures of a fixed similarity type. The idea underlying this definition is that to define a theory one needs to define its *scope*, i.e. the class of structures that the theory is supposed to describe, rather than define its *content*, i.e. the set of laws of which the theory consists.

While the deductive notion of a theory has a rather long history, the set-theoretic one is of a quite recent origin. In an explicit way it was defined by Patrick Suppes [7]. The *non-statement view*, as the idea to identify theories with their scopes has been called, corresponds to the practice common among mathematicians that consists in defining mathematical theories by defining structures to be examined within them. Thus, for instance, they define the theory of Boolean algebras by defining what is a Boolean algebra, they define the theory of probability by defining what is to be understood by a probability space, etc.

As should be clear from the above examples, to define a theory in the set-theoretic sense is to define a set-theoretic predicate P (Boolean algebra, probability space, etc.) whose extension is the set of structures meant to be the theory in question But then the clauses of the definition of P form an axiom system for a theory T in the deductive sense, the language in which they are stated being the language of T. The theory T both defines P and is defined by P. Thus, each theory in the set-theoretic sense is reducible, if given explicitly, to a theory in the deductive sense (see also [5]), and we may safely restrict ourselves to using the deductive notion of a theory only. Actually, in what follows I shall use the term 'theory' in the informal sense

P. I. Bystrov and V. N. Sadovsky (eds.), Philosophical Logic and Logical Philosophy, 83–89.

only. Formalized counterparts of informal theories will be referred to as deductive systems or deductive theories.

With the advent of mathematical logic, mathematics has been successfully studied by studying theories of which it consists. It was only natural to expect that the logical approach would turn useful in dealing with empirical science. Philosophy of science based on this expectation was logical empiricism (neopositivism); for philosophers of this persuasion (R. Carnap, K. Ajdukiewicz, K. Popper, H. Richenbach, C. G. Hempel and many others) logic was the basic tool for theorizing about science, the latter being viewed as a collection of deductive systems. Thus a deductive theory was taken to be the central unit of the analysis.

Kuhn's criticism of a logical conception of science [2] marks the beginning of the historical turn in the philosophy of science. Rather than focus attention on those activities of science which can be analysed entirely in logical terms, Kuhn and other post-positivist theorists of science stress the significance of psychological and sociological factors for rendering our understanding of science and its evolution deeper and more coherent. An increased emphasis is put on the fact that science is a social construct; one cannot do science without an appropriate social training, part of which consists in accepting views and norms shared by scientific communities.

It goes without saying that psychological and sociological phenomena relevant to scientific enterprise may be of considerable interest for the historian of science. Also it is fairly obvious that they play an important role in science formation. Individual interests, conservatism of professional groups, urgency of certain social needs are some obvious examples of factors which may influence the growth of science. Still, one may doubt whether philosophy of science should be combined with sociology and psychology. Reasoning is a psychological phenomenon, still logic became a full-fledged discipline only after it had been separated from psychology. Communication is a social phenomenon, but theoretic linguistics, being a rather prosperous part of linguistic studies, has nothing to do with sociology nor its methods.

Quite a different thing is that the neopositivist theory of science has been unable to provide a sufficiently accurate account of the structure of science nor its dynamics. But this failure seems to have little to do with the fact that positivists ignored social aspects of scientific activity.

2 Illusions of Logical Reconstruction

In order to use logical techniques in studying science one has to start with formalizing that part of it which has to be studied. In particular, theories to be examined have to be stated in the form of deductive systems. The procedure of transforming informal theories into their formal counterparts is known as *logical reconstruction*. The usefulness of this technique consists in that it allows its user to define the 'mathematical skeleton' of the reconstructed theory in a precise manner, i.e. to present the theory in the form of a set of mathematical theorems. Unfortunately to reconstruct the mathematical skeleton is not the same as to reconstruct the theory. Still worse, unless criteria of adequacy are dramatically relaxed, any adequate formal reconstruction of an empirical theory is not available. The main reasons why it is not are the following:

There are empirical quantities which cannot be adequately characterized within a single theory

An example of a quantity whose properties, even formal, cannot be adequately characterized within one theory is energy. For no part of physics, is energy a specific magnitude of it. Rather it appears in all physical theories and all of them contribute to some extent to defining it. Consequently, while mathematical theories provide a complete characteristic of their primitive terms, this is not true of empirical ones.

There are empirical principles which cannot be adequately stated within one theory

An example of a principle which is valid in all domains of physics, actually in all domains of empirical science, and thus cannot be adequately stated within any single theory, is the principle of invariancy of physical laws under Lorenzian transformations.

Logical reconstruction separates theories from their methodologies

It is often argued that theories should not be confused with their methodologies. But methodology is not merely a set of norms that govern the behaviour of people doing science. It always contains a descriptive part. Consider the following example. Any application of Particle Mechanics, PM, to a three-dimensional empirical body involves the procedure of dividing the body into parts small enough to be treated as '0-dimensional', thus as particles in the sense of PM. The procedure is implicitly based on an empirical claim which could be fairly easily and fairly accurately stated. So, why do physicists not state it? For two reasons:

First, because they avoid introducing to theoretic considerations any notions whose properties are of empirical rather than of a mathematical (formal) nature. The notion of a physical body is of this kind. If it were included into particle mechanics in addition to that of a particle (a particle point), the mathematical structure of theory would be blurred.

Second, the number of empirical claims which form a part of methodology, but do not belong to the content of the theory is usually enormous. Moreover, any newly discovered area of application extends the descriptive part of methodology. Consequently the body of descriptive methodological claims is continually expanding and thus is practically unlimited. By no means can it be reduced to a few laws stated in an appropriate mathematical language. This explains why it is ill-advised to treat general theoretic principles on a par with descriptive methodological claims, thus treat the former and the latter as being of the same epistemic status.

Perhaps it is worth noticing also that a theory separated from its methodology is just a piece of mathematics and can hardly be viewed as an adequate reconstruction of its informal counterpart.

Besides well established claims theories may involve some vague and often conflicting hypotheses

The latter may happen to convey ideas of considerable significance for further developments. Any logical reconstruction of a theory cuts across its hypothetical part, if not removing this part entirely. Consequently the resulting deductive system may happen to be a drastically oversimplified version of the reconstructed unit.

The deductive apparatus of informal theories does not reduce to any system of well codified rules of inference

Whether or not a specific mathematical transformation (differentiation, using the initial terms of the Taylor series, etc.) can be applied in a reasoning carried out within an informal theory depends on some factual considerations. On the other hand, the very idea of a deductive system presupposes a rigid formalization of the deductive apparatus.

3 Theories vs. Models

Theories in the informal sense of the word are not clear-cut separated units of science. Neither their contents, nor their scope is well defined. One should not expect that every descriptive term that an empirical theory involves is fully characterized by the principal laws of the theory, and one should not expect that an augment carried out within a theory is entirely based on its laws and the mathematical formalism characteristic of it.

A theory reconstructed in the form of a deductive system becomes a rigorously defined unit: a calculus whose elements are governed by given in advance transformation rules. A somewhat metaphorical term 'mathematical skeleton' seems to express relatively well the relation that holds between a theory and its formal counterpart being the product of logical reconstruction.

Suppose that a deductive system is furnished with an empirical interpretation. Under the positivist conception of science, an empirical interpretation consists in reducing theoretical terms to observational ones. Different definitions of how this reduction is to be carried out were proposed at different periods of time. There are not many theorists of science who would still be ready to subscribe to the idea of an observational language and the corresponding idea of the observational-theoretical dichotomy. But the reduction of theoretical terms to observational ones is but one of many ways in which deductive systems can be supplied with an empirical interpretation.

The simplest and the most natural way to interpret empirically mathematical statements is to relate mathematical quantities they involve to appropriately selected measurement procedures combined with techniques of processing empirical data. An interpretation of this kind may be called *operational*. It would take time and effort to define the idea of the operational interpretation in a coherent and precise way. On the other hand, the operational interpretation is the one which is applied in the real life. The most common way to apply theoretic quantities to empirical phenomena consists in defining how to measure them. One should not interpret this, rather obvious, observation as an access to Bridgman's operationalism or operationalistic philosophy of any kind. The doctrine of operationalism is based on the belief that empirical quantities are defined by measurement procedures. The operational interpretation of theoretic quantities can be applied without accepting this view.

An operationally interpreted deductive system is a mathematical model of an empirical phenomenon (or phenomena) from the scope of the theory. The substantial difference between a model and a theory consists in that the former is not meant to provide any explanation of the phenomenon to which it refers. its role is purely instrumental. Solving questions concerning a model (for instance, finding roots f the equations of which the model consists)is a way to solve questions concerning the modelled phenomenon. The justification of a model is solely and precisely that it is expected to work.

Theories play quite a different role. Their task is to provide us with a conception of the reality to which the theory applies. Since they serve us to understand the reality, they cannot be separated from ontological beliefs underlying them, methodological principles that govern their applications and even from the intuitive knowledge of practitioners. Historically minded philosophers of science have rightly emphasised the significance of these supposedly extra-theoretic factors. One of the conclusions of their analyses was that empirical theories should be studied as parts of larger units such as *paradigms* [2], *research programs* [3], *global theories* [1], *research tradition* [4] and *theory nets* [6]. the idea that the role of theories is to allow us to understand the reality, while the role of mathematical models related to them is to allow us to cope with the reality is entirely alien to positivist philosophy of science. Understanding is not a category which a positivistically minded philosopher is ready to have in his or her vocabulary. No wonder then, that the doctrine of logical empiricism identifies theories with empirically interpreted deductive systems.

Our analysis of the interrelations between theories and models has been incomplete. Logical reconstruction is a way in which one arrives from a theory to a model. But there are not too many empirical theories which have only one intended empirical interpretation. Typically, each application of a theory requires a separate interpretation which, when implemented, results in defining a model suitable to this application. In this sense models can be viewed as applications of theories—for instance the model of oceanic tides formed with the help of Newtonian Particle Mechanics is a paradigmatic application of the latter.

The following should be stressed. Interrelations between theories and models cannot be adequately described in logical terms. In order to form a model of a specific phenomenon to which a theory is applicable, besides the laws of the theory in question, one usually uses some additional, often intuitive knowledge. Thus forming models, and consequently interpreting theories, is a creative activity which cannot be performed in a routine way.

While positivist philosophers of science focused their attention on empirically interpreted deductive systems, thus on models, post-positivist philosophers focus their attention on theories, neglecting the fact that we need theories in order to form models. The chief task of science is to provide knowledge allowing us to control our environment and to fulfil our needs. Since the knowledge that directly allows us to cope with practical problems is given in the form of models of empirical phenomena rather than in the form of theories, the role a specific part plays in the whole body of science as well as the role it plays in the process of growth of science can be adequately judged only by evaluating its usefulness for producing models. The phenomenon of science cannot be adequately understood, if the role of models is neglected. In fact, models should be treated as the central unit of any theorising about science.

In order to illustrate this point, let me briefly mention one issue only. The widely discussed problem of cumulativity of science cannot be adequately defined without appealing to the notion of a model. Science is cumulative in the sense that although good models may be replaced by better ones, it does not mean that the former are to be discarded, declared to be false, or that using them becomes inappropriate in any sense. Thus one may see science as a continuous process of producing more and more models to handle practical problems. Viewed in this way science is definitely a cumulative enterprise.

Theories, ways in which we understand reality, perspectives under which we see things surrounding us are in a constant flux. They are cumulative in no reasonable sense, for forming a new theory by no means consists of adding new elements to the old ones.

4 A Concluding Remark

In order to state things above discussed in a rigorous language, thus to go beyond some preliminary rough ideas, one should start with an analysis of the notion of a model. In fact, this is the notion of a model not that of a theory which is central for the analysis I presented here in a sketchy way. for an attempt to define this notion, see [8].

Polish Academy of Sciences, Warsaw

References

[1] P. Feyerabend. *Against Method*, New Left Books, London, 1975.

[2] T. S. Kuhn. *The Structure of Scientific Revolutions*, 2nd ed., enlarged, University of Chicago Press, Chicago, 1970.

[3] I. Lakatos. *The Methodology of Scientific Research Programmes*, Cambridge University Press, Cambridge, 1978.

[4] L. Laudan. *Progress and Its Problems*, University of California Press, Berkeley, CA, 1977.

[5] D. Pearce and V. Rantala. A logical study of the correspondence relation. *Journal of Philosophical Logic*, **13**, 1984.

[6] W. Stegmüller. *The Structualist view of Theories*, Springer-Verlag, Berlin, 1979.

[7] P. Suppes. *Introduction to Logic*, Van Nostrand, New York, 1957.

[8] R. Wójcicki and E. Kałuszyńska. *Theories, Models, Phenomena*, paper to be read at International Symposium on Structures in Mathematical Theories, San Sebastian, Spain, 1990.

7

Structures, Suppes Predicates, and Boolean-Valued Models in Physics

N. C. A. da Costa and F. A. Doria

1 Introduction

An old and important question concerning physical theories has to do with their axiomatization [47]. The sixth problem in Hilbert's celebrated list of mathematical problems deals with its desirable (or ideal) contours [31]:

> *The Mathematical Treatment of the Axioms of Physics.*
>
> The investigations on the foundations of geometry suggest the problem: *to treat in the same manner, by means of axioms, those physical sciences in which mathematics plays an important part; in the first rank are the theory of probability and mechanics.*
>
> As to the axioms of the theory of probabilities, it seems to me to be desirable that their logical investigation be accompanied by a rigorous and satisfactory development of the method of mean values in mathematical physics, and in particular in the kinetic theory of gases.
>
> Important investigations by physicists on the foundations of mechanics are at hand; I refer to the writings of Mach..., Hertz..., Boltzmann..., and Volkman... It is therefore very desirable that the discussion of the foundations of mechanics be taken up by mathematicians also. Thus Boltzmann's work on the principles of mechanics suggests the problem of developing mathematically the limiting processes, those merely indicated, which lead from the atomistic view to the laws of continua. Conversely one might try to derive the laws of motion of rigid bodies by a limiting process from a system of axioms depending upon the idea of continuously varying conditions on a material filling all space continuously, these conditions being defined by parameters. For the question as to the equivalence of different systems of axioms is always of great theoretical interest.

P. I. Bystrov and V. N. Sadovsky (eds.), Philosophical Logic and Logical Philosophy, 91–118.

> If geometry is to serve as a model for the treatment of physical axioms, we shall try first by a small number of axioms to include as large a class as possible of physical phenomena, and then by adjoining new axioms to arrive gradually at the more special theories. At the same time Lie's principle of subdivision can perhaps be derived from the profound theory of infinite transformation groups. The mathematician will have also to take account not only of those theories coming near to reality, but also, as in geometry, of all logically possible theories. We must be always alert to obtain a complete survey of all conclusions derivable from the system of axioms assumed.
>
> Further, the mathematician has the duty to test exactly in each instance whether the new axioms are compatible with the previous ones. The physicist, as his theories develop, often finds himself forced by the results of his experiments to make new hypotheses, while he depends, with respect to the compatibility of the new hypotheses with the old axioms, solely upon these experiments or upon a certain physical intuition, a practice which in the rigorously logical building up of a theory is not admissible. The desired proof of the compatibility of all assumptions seems to me also of importance, because the effort to obtain such a proof always forces us most effectively to an exact formulation of the axioms.

At present, two basic procedures to axiomatize a theory are the following: to make explicit its underlying mathematical structure, in the sense of Bourbaki, and to construct the Suppes predicate that formalizes it. Both methods are, in effect, essentially equivalent (see [5] [10] [43] [44] [45]). Mathematical structures and the Suppes set-theoretical predicates, of course, constitute set-theoretic constructs.

With the help of ideas and methods originated by the works of Gödel and Cohen, we can build infinitely many models of set theory (cf. for example [6]). In those models some significant principles, such as the Axiom of Choice, the Generalized Continuum Hypothesis, and Martin's Axiom, may or not be valid. It is therefore natural to enquire whether the use of one or another of those models of set theory have consequences for the general theory of mathematical structures; in particular, whether the employment of a certain kind of model has consequences for the axiomatization of physical theories. That question can also be seen as motivated by Hilbert's discussion of the axiomatic method in physics, quoted above, when he says that we have to take into consideration

> [...] not only [...] those theories coming near reality, but also, as in geometry, [...] all logically possible theories.

Results in that direction have already been obtained, for example, by Wright [49]. That author investigates some aspects of the development of Solovay's mathematics [40], i.e. of mathematics based on a special model of set theory. Among other interesting results, it is shown that the theory of Hilbert spaces based on Solovay's model of set theory does not coincide with the classical version. Previous explorations of the same theme can be found in [7] [8] [35]. Work by the authors on Suppes predicates starts with its application to general relativity [21]; a more intuitive exploration of Suppes predicates as a tool in the formalization of physical theories can be found in [11]; metamathematical incompleteness of physics, also based on Suppes predicates, is given in [12] [13] [23] [24] [19].

In the present paper we give some examples, chosen within established physical theories, which clearly indicate that the underlying models of set theory, involved in our axiomatization of physics, have not only mathematical relevance, but also import from the physical point of view (in that respect our position is different from that of [7] [8], since that author tries to exclude set-theoretic models by imposing on them physically-determined conditions, which the models may or may not satisfy).

For the sake of completeness we delineate the concepts of species of mathematical structures and of Suppes predicates before the discussion of our examples.

2 The Bourbaki Structures

We now informally describe Bourbaki's concept of a species of structures. We will be working within a standard system of set theory, such as, say, the Zermelo–Fraenkel system [6] [34].

According to Bourbaki [5], Chapter 4, we define an **echelon construction scheme** to be a finite sequence

$$\langle c_1, c_2, \ldots, c_m \rangle$$

of ordered pairs of natural numbers $c_i = \langle a_i, b_i \rangle$, $1 \leq i \leq m$, such that:

1. If $b_i = 0$, then $1 \leq a_i \leq i - 1$;
2. If $a_i \neq 0$ and $b_i \neq 0$, then $1 \leq a_i \leq i - 1$ and $1 \leq b_i \leq i - 1$.

Therefore we have that $c_1 = \langle 0, b_1 \rangle$ and $b_1 > 0$. If n is the greatest value of the second members of the pairs

$$c_1, c_2, \ldots, c_m,$$

then that ordered sequence (which we note S) is called an **echelon construction scheme of order** n.

If

$$S = \langle c_1, c_2, \ldots, c_m \rangle$$

is an echelon construction scheme of order n and if $X_1, X_2, \ldots, X_n$ are n sets, then **an echelon construction of scheme** S **on** $X_1, X_2, \ldots, X_n$ is the sequence $\langle A_1, A_2, \ldots, A_m \rangle$ defined as follows:

1. If $c_i = \langle 0, b_i \rangle$, then $A_i = X_{b_i}$;

2. If $c_i = \langle a_i, 0 \rangle$, then $A_i = \mathcal{P}(X_{b_i})$, where $\mathcal{P}$ denotes the power set operation on X_{b_i};

3. $c_i = \langle a_i, b_i \rangle$, where $a_i \neq 0$ and $b_i \neq 0$, then $A_i = X_{a_i} \times X_{b_i}$.

The set A_m is an **echelon of scheme S on the base sets** $X_1, X_2 \ldots, X_n$; we note:

$$A_m = S(X_1, X_2, \ldots, X_n).$$

Let us now be given:

1. An echelon construction scheme $S = \langle c_1, c_2, \ldots, c_m \rangle$ on n sets;

2. Two finite sequences of sets $\langle X_1, X_2, \ldots, X_n \rangle$ and $\langle Y_1, Y_2, \ldots, Y_n \rangle$;

3. A sequence of functions

$$f_i : X_i \to Y_i, \ 1 \leq i \leq n;$$

4. The echelon construction schemes of order n on $X_1, X_2, \ldots, X_n$ and on Y_1, $Y_2, \ldots, Y_n$, $\langle A_1, A_2, \ldots, A_m \rangle$ and $\langle A'_1, A'_2, \ldots, A'_m \rangle$, respectively.

We can then define in a step-by-step way a sequence $\langle g_1, g_2, \ldots, g_m \rangle$, where $g_i : A_i \to A'_i$, by the conditions:

1. If $c_i = \langle 0, b_i \rangle$, then $A_i = X_{b_i}$ and $A'_i = Y_{b_i}$, we put $g_i = f_{b_i}$;

2. If $c_i = \langle a_i, 0 \rangle$, then we have that $A_i = \mathcal{P}(A_{a_i})$, where $\mathcal{P}$ denotes the power set operation, and $A'_i = \mathcal{P}(A'_{a_i})$, we put $g_i = \hat{g}_{a_i}$, where $\hat{g}_{a_i}$ is the canonical extension of g_{a_i} to $\mathcal{P}(A_{a_i})$ (and $\mathcal{P}(A'_{a_i})$, respectively);

3. If $c_i = \langle a_i, b_i \rangle$, where $a_i \neq 0$ and $b_i \neq 0$, then $A_i = A_{a_i} \times A_{b_i}$, $A'_i = A'_{a_i} \times A'_{b_i}$, and we put $g_i = g_{a_i} \times g_{b_i}$, where $g_{a_i} \times g_{b_i}$ is the canonical extension of g_{a_i} and g_{b_i} to $A_{a_i} \times A_{b_i}$.

We denote g_m by $\langle f_1, f_2, \ldots, f_m \rangle^S$.

We call **typifications** expressions of the form

$$s \in S(X_1, X_2, \ldots, X_m).$$

In a typification some of the sets $X_1, X_2, \ldots, X_m$ are said to be **principal sets** while the remainder consist of **auxiliary sets**. There are always principal sets, while there may be no auxiliary sets. Therefore, we may represent a typification as follows:

$$s \in S(X_1, X_2, \ldots, X_p, A_1, \ldots, A_q),$$

where the X's are the principal sets, while the A's are the auxiliary sets; $p+q=m$. Usually the A's are sets whose relevant properties have been previously studied.

Let $F(X_1, \ldots, X_p, A_1, \ldots, A_q, s)$ be a formula of set theory. Let $f_i : X_i \to X'_i$, $1 \leq i \leq p$, be a sequence of functions and $I_j : A_j \to A_j$ be the identity mapping of A_j, $1 \leq j \leq q$. We say that F is a **transportable formula** with respect to the typification $s \in S(X_1, \ldots, X_p, A_1, \ldots, A_q)$ if we have that:

$$F(X_1, \ldots, X_p, A_1, \ldots, A_q, s) \leftrightarrow F(X'_1, \ldots, X'_p, A_1, \ldots, A_q, s'),$$

where $s' = \langle f_1, \ldots, f_p, I_1, \ldots, I_q \rangle^S$, and $\leftrightarrow$ denotes material equivalence.

Then, for Bourbaki, a **species of structures** is the class defined by a predicate $P(s)$ of the form:

$$\begin{aligned} P(s) \quad \leftrightarrow \quad & \exists X_1 \ldots \exists X_p (s \in S(X_1, \ldots, X_p, A_1, \ldots, A_q) \wedge \\ & \wedge F(X_1, \ldots, X_p, A_1, \ldots, A_q, s), \end{aligned}$$

where $s \in S(X_1, \ldots, X_p, A_1, \ldots, A_q)$ is the typification of the species of structures and F is its axiom. F is supposed to be transportable; as usual, $\wedge$ denotes conjunction.

Any s that satisfies the predicate P is said to be **a structure of species** P. The usual mathematical structures are particular instantiations of the preceding definition, as seen in [5], Chapter 4. Some species of structures, namely those of field, group, or topological space, do not require auxiliary base sets, while others, such as those of real vectorspace and complex manifold need auxiliary sets (the real field and the complex field, respectively).

The present construction summarizes the one given in full detail in [10]. For a more intuitive approach, see [11] [21].

It should be noted that Bourbaki's treatment of the concept of species of structures (and also of the notion of structure) presents some drawbacks, as it is purely syntactical. However that difficulty can be easily overcome with the help of methods from model theory [10].

In most cases the typification

$$s \in S(X_1, \ldots, X_p, A_1, \ldots, A_q)$$

of a species of structures may be decomposed as a Cartesian product

$$S_1(X_1, \ldots, X_p, A_1, \ldots, A_q) \times \ldots \times S_k(X_1, \ldots, X_p, A_1, \ldots, A_q),$$

so that s reduces to a k-tuple (note that in the scheme S_n, $1 \leq n \leq k$, some of the sets $X_1, \ldots, A_1$ may appear vacuously).

The concepts of isomorphism between structures, transport of structures, equivalent species of structures, initial structures, final structures, and so on, are defined in [5], Chapter 4.

3 Suppes Predicates

We begin with an example. Let us consider a typical mathematical concept; for instance, that of semigroup. A semigroup is an ordered pair $\langle X, f\rangle$, where X is a nonempty set and f is a binary associative operation on X. Ordered pairs, binary operations, associative operations, and nonempty sets can be easily defined in set theory, and we can reformulate our definition for semigroup as follows:

$$G(x) \leftrightarrow \exists X \exists f(x = \langle X, f\rangle \wedge P(X) \wedge Q(f, X)),$$

where $P(X)$ stands for 'X is a nonempty set', and $Q(f, X)$ means 'f is an associative binary operation on X'.

The predicate $G(x)$ means, 'x is a semigroup', and the expression at the right side of the equivalence connective may then be supposed to be formulated in the language of set theory, that is, $G(x)$ is a set-theoretical predicate. In general, any kind of mathematical structure can be so defined. We may legitimately assert with Suppes that to axiomatize the theory of a certain kind of structure consists in defining a set-theoretic predicate. As Suppes writes,

> In the first place, it may be well to say something more about the slogan 'to axiomatize a theory is to define a set-theoretical predicate.' It may not be entirely clear what is meant by the phrase 'set-theoretical predicate.' Such a predicate is simply a predicate that can be defined within set theory in a completely formal way. For a set theory based only on the primitive predicate of membership, this means that ultimately any set-theoretical predicate can be defined solely in terms of membership. Any standard mathematical notions will be used freely in the definiens of such a set-theoretical definition of a predicate, for we shall assume that these standard notions have already been fully developed and formalized. Such a development of mathematics is in fact to be found in the many-volumed treatise written under the collective pseudonym of Bourbaki. Whenever a previous set-theoretical formalization of an *empirical* theory is assumed, this shall be explicitly so indicated [. . .].
>
> This last remark suggests that there may be some systematic difference between a set-theoretical definition embodying concepts of pure mathematics and one involving concepts of the empirical sciences. I do not mean to suggest anything of the sort. It is one of the theses of this book that there is no theoretical way of drawing a sharp distinction between a piece of pure mathematics and a piece of theoretical science. The set-theoretical definitions of the theory of mechanics, the theory of thermodynamics, the theory of learning, to give three rather disparate examples, are on all fours with the definitions of the purely mathematical theories of groups, rings, fields, etc. From the philosophical standpoint there is no sharp distinction between pure and applied mathematics, in

spite of much talk on the contrary ([44], Chapter 2, pp. 29–30; see also [45]).

However we notice that the concept of a Suppes predicate is slightly more general than that of a Bourbaki species of structures. For example,

$$K(x) \leftrightarrow x = \emptyset,$$

constitutes a Suppes predicate, but does not define a (nontrivial) species of structures. Nevertheless, for our present goals, we may identify both concepts and consider the *models* of Suppes predicates as the same as the structures of a convenient species [10].

A Suppes predicate determines the *theory* of the structures which satisfy it; it also determines which are its models. We thus conclude that to axiomatize (or to formalize) a theory—of mathematics or of the empirical sciences—means simply that we should make explicit its Suppes predicate. Section 5 of the present paper discusses three examples of such an axiomatization, and its model-theoretic consequences. We examine topological spaces and their behavior under expanded models via a Boolean extension; that example includes the effect of such an expansion on the boundary conditions for field equations in classical physics. We then examine Bernouilli shifts and their entropies (which may even be zeroed, while the original set of Bernouilli sequences may even cease to be a measurable set). Finally, we consider the case of set-theoretically generic electromagnetic fields.

In all cases we make explicit the corresponding Suppes predicate, which then leads to the associated Bourbaki species of structures. For our recent work on related questions see [11] [12] [13] [23] [21] [24] [19].

4 Axiomatization

Our exposition shows that the axiomatic method went through a profound evolution since its appearance in Euclid's *Elements*. Today it goes beyond the merely syntactical dimension of language and involves a semantical counterpart, i. e., it incorporates model- and set-theoretical ideas (see [10]). In particular, the very concept of a mathematical structure, which is the current basic tool in the axiomatization process, has also suffered a deep change in the course of history, a fact that indicates that even the main concepts that one deals with in the foundations of mathematics do also progress.

Since our way of getting axiomatized theories is through mathematical structures embedded into a given set theory, we have that set theory is a decisive construct in our approach for the axiomatic method. In the last decades, the work of Gödel and Cohen has given rise to a conceptual revolution in set theory; today we know lots of different models for it. As an example, in Solovay's model (which satisfies the axioms of Zermelo–Fraenkel set theory + the axiom of dependent choices [33] [40]), any set of real numbers is Lebesgue-measurable and satisfies the Baire property. Moreover, the theory of Hilbert spaces is drastically modified, as Wright has shown that all operators on a Hilbert space are bounded [49]. If we take into account those facts, we see that the model of set theory we use as the foundation for a certain species of structures has decisive influence on its development.

The axiomatization of physical theories starts from the mathematical task of defining convenient mathematical structures as the setting where we develop our physical theory. As we show in the examples discussed in the next Section, the background model will have a decisive influence on what kind of theory we get, even when that model satisfies a 'tame' theory such as Zermelo–Fraenkel plus the full axiom of choice and, say, a technical axiom such as Martin's axiom, which essentially acts as a regularizing tool that allows us to have 'nice,' 'well-behaved' properties in the model. Sometimes even the system Zermelo Fraenkel + full axiom of choice + generalized continuum hypothesis + there are nonconstructive sets allows us to obtain results which differ from those that we obtain when we impose Gödel's axiom of constructibility. On similar questions see [7] [8] [35].

5 Examples

The set-theoretical notions that we require in the present Section can be found in [6] [33] [34]; our notation follows [6]. We suppose that all the models discussed here satisfy ZFC, that is, Zermelo–Fraenkel set theory (with the axiom of regularity) plus the full axiom of choice. Therefore our results are compatible with the whole of 'classical' mathematics, which is identified to the set of theorems that can be deduced from the ZFC axioms. However our examples are centered on the consequences of adding 'inessential' (or 'non-classical') axioms to ZFC, such as Gödel's axiom of constructibility, the negation of the continuum hypothesis, or Martin's axiom.

As it is usual, CH abbreviates 'the continuum hypothesis', and GCH is 'the generalized continuum hypothesis'. MA stands for 'Martin's axiom', while MA(κ) is 'Martin's axiom up to a cardinal κ such that $\aleph_0 \leq \kappa < 2^{\aleph_0}$'. If X is a set, then $|X|$ denotes the set's cardinality. ω_0 is the set of natural numbers, Z are the integers, R are the reals and C are the complex numbers; Ir $\subset [0,1]$ are the irrationals in the unit interval, while **L** is Gödel's constructible universe. Further notation will be described when needed.

Preliminary Results

We now quote without proof a few results that will be required in the examples. Those that deal with MA are proved in [33] [34] [39]; the remaining results are found in [33].

Proposition 5.1 If the theory ZFC + 'CH is false' + 'MA is true' is consistent, then we can derive from its axioms that if X is a topological space with a countable base, and if κ is a cardinal such that $\aleph_0 \leq \kappa < 2^{\aleph_0}$, then the union of $\leq \kappa$ meager sets in X is meager.

PROOF. See [39], Theorem 7. Kunen's proof [34] of a particular case ($X = \mathsf{R}$, where R denotes the reals) can be easily modified to give the more general result. ∎

Proposition 5.2 If the theory ZFC + 'CH is false' + 'MA is true' is consistent, then we can derive from its axioms that, given a family of subsets $X_\alpha \subset \mathsf{R}$ of the reals, $\alpha < \kappa$ and $\aleph_0 \leq \kappa < 2^{\aleph_0}$, if the Lebesgue measure of each one of the X_α is zero, then $\cup_{\alpha<\kappa} X_\alpha$ has measure zero.

PROOF. See the references. ∎

The next result is somewhat complementary to the preceding propositions:

Proposition 5.3 If the theory ZFC + $\mathbf{V} = \mathbf{L}$ is consistent, then:

1. There is a model for ZFC + $\mathbf{V} \neq \mathbf{L}$ such that the assertion 'the set of constructible reals $\widehat{\mathsf{R}} = \mathsf{R} \cap \mathbf{L}$ isn't Lebesgue measurable' is true in that model.
2. There is a model for ZFC + $\mathbf{V} \neq \mathbf{L}$ such that the assertion 'the set of constructible reals $\widehat{\mathsf{R}}$ does not have the property of Baire' is true in that model.

PROOF. See [33], p. 567. ∎

Remark 5.4 If we wish to discuss the related problem of the measure and category properties of a subset $X \subset \mathsf{R}$ whose cardinality is strictly smaller that that of the continuum within the ZFC axioms supposed consistent, we can only derive from them on that question the following two assertions:

> For a subset $X \subset \mathsf{R}$ such that its cardinality $|X| < 2^{\aleph_0}$, if X is Lebesgue measurable then it has zero Lebesgue measure.
>
> For a subset $X \subset \mathsf{R}$ such that its cardinality $|X| < 2^{\aleph_0}$, if X has the property of Baire then it is a meager set.

See [37]. And from ZFC + 'there is a nonconstructible set', we can only prove that either $\widehat{\mathsf{R}}$ is of zero Lebesgue measure or it isn't measurable; and again either $\widehat{\mathsf{R}}$ is a meager set or doesn't have the property of Baire. For a reference see [33]. If we add to that theory the denial of CH plus MA, one tilts the scale in favor of the constructible reals being a meager set of zero measure; the other models described in the reference leave them non-measurable and without the property of Baire.

Recall that a *Polish space* is a topological space homeomorphic to a complete and separable metric space. Let us be given $\omega_0^{\omega_0}$ where ω_0 has the discrete topology and the product space the corresponding induced topology; as there is a natural homeomorphism $\omega_0^{\omega_0} \cong \mathrm{Ir} \subset [0,1]$ with the irrationals in the unit interval, induced by the continuous fraction representation of irrationals, then $\omega_0^{\omega_0}$ can be given a measure induced from that of Ir, and compatible with its natural product topology. Then $\omega_0^{\omega_0}$ is universal for all Polish spaces, that is,

Lemma 5.5 If X is a Polish space, then there is an open and onto projection $\varphi : \omega_0^{\omega_0} \to X$.

PROOF. See [3] [33]. Sometimes φ can be even algorithmically defined, so that computable functions in the covering space are mapped over Pour-El computable objects in X. See on that [25] [38].

As a consequence, we have in ZFC that:

Corollary 5.6 If X is Polish, then $|X| \leq 2^{\aleph_0}$; if X is Polish and uncountable, then $|X| = 2^{\aleph_0}$.

PROOF. The first statement is a consequence of Lemma 5.5; the second statement results from the preceding one and from the fact that an uncountable Baire space such as X has precisely $2^{\aleph_0}$ as its cardinality.

Now suppose that we are handling objects from the set-theoretic universe $\mathbf{L}$. If X is Polish, then clearly $\mathbf{L} \models |X| \leq \aleph_1$. Let $C \in \mathbf{L}$ be a c.c.c. complete Boolean algebra [6] [34]. If $\mathbf{L}^C$ is the corresponding Boolean extension, then, if $\mathbf{L} \models$ 'X is Polish', we have that:

Proposition 5.7 $\mathbf{L}^C \models |\widehat{X}| \leq \aleph_{\hat{1}}$.

PROOF. We restrict our attention to the uncountable case, since the other cases are dealt with by absoluteness. Since C is c.c.c., $\mathbf{L} \models |X| = |\aleph_1| = \aleph_1$ if and only if $\mathbf{L}^C \models |\widehat{X}| = \widehat{\aleph}_1$. Again due to c.c.c., we have that $\mathbf{L}^C \models \widehat{\aleph}_1 = \aleph_{\hat{1}}$.

Corollary 5.8 If $\mathbf{L} \models$ 'X is a Polish space', then, if $\mathbf{L}^C \models$ ZFC + 'MA is true' + 'CH is false', we have that $\widehat{X}$ is meager in X in the Polish space topology.

PROOF. From Proposition 5.1 and Corollary 5.6.

Let $B, B' \in \mathbf{L}$ be c.c.c. complete Boolean algebras so that the extensions $\mathbf{L}^B$ and $\mathbf{L}^{B'}$ satisfy respectively the two conditions in Proposition 5.3. Then:

Corollary 5.9

1. $\mathbf{L}^B \models$ 'If $X \subseteq [0,1]$ is Polish in the induced topology, then $\widehat{X} \subset X$ isn't Lebesgue measurable'.
2. $\mathbf{L}^{B'} \models$ '$\widehat{X}$ doesn't have the Baire property in X'.

PROOF. From 5.6 and Proposition 5.3.

Our interest in Polish spaces lies in that they appear everywhere in theoretical physics. Spacetime manifolds are Polish, as well as the configuration and phase spaces that appear in mechanics; moreover, most function spaces ('field spaces') that appear in physics are also Polish.

We are now ready to discuss our examples.

A topological space may get new open sets when we change the set-theoretical model.

A topological space is an ordered pair $\langle X, \tau\rangle$, where X is a nonempty set and $\tau \subseteq \mathcal{P}(X)$ is a distinguished subset; as usual, $\mathcal{P}$ denotes the power set operation. Furthermore we impose that both the empty set $\emptyset$ and X belong to τ, and that τ be closed with reference to arbitrary unions and finite intersections. We can also define τ out of a characteristic function on $\mathcal{P}$ subject to restrictions equivalent to the ones above. The corresponding Suppes predicate is:

$$T(x) \leftrightarrow \exists X \exists \tau (x = \langle X, \tau\rangle \wedge \tau \subseteq \mathcal{P}(x) \wedge \emptyset \in \tau \wedge X \in \tau \wedge$$
$$\wedge\ \tau \text{ is closed with respect to arbitrary unions } \wedge$$
$$\wedge\ \tau \text{ is closed w.r.t. finite intersections}),$$

where the predicate $T(x)$ means 'x is a topological space' and the expression to the right of the material equivalence connective is taken within the language of set theory.

Now most physically interesting properties are *local* properties, that is, properties which are valid over an open set (a sufficiently small open set) on spacetime. Consider just the usual idealization for a laboratory situation: when one performs an experiment in a laboratory, everything happens within the product $L \times t$, where L denotes laboratory space and t is the time interval that encompasses our experiment. Some local properties are valid for 'nice' open sets (say, homotopically trivial open sets or even open domains homeomorphic to an open ball). Other properties may be valid for arbitrary open sets, which may have highly complex topological properties—for example, the domains over which a holonomy group in classical gauge field theory changes its dimension (see [27]).

Suppose that $\langle X, \tau\rangle$ is a Hausdorff space with a countable base; the cardinality $|\tau| = 2^{\aleph_0}$. Let $\mathbf{L}$ be the constructible universe and let $B = \mathrm{RO}(2^{\aleph_0})$ be the regular open algebra of the product $2^{\aleph_0}$ (see [6]). Then $\widehat{\tau}$ denotes as usual the image of $\tau \in \mathbf{L}$ within $\mathbf{L}^B$ through the standard embedding of the corresponding universes, while τ^B is the topology of X in $\mathbf{L}^B$. Furthermore we ask that X (in $\mathbf{L}$) be given an open cover

$$\mathcal{X} = \{U_i : i \in \omega_0\}$$

subject to the conditions:

1. $\cup_{i \in \omega_0} U_i = X$.
2. For $i \neq j$, it is false that there are k so that $U_k \subseteq U_i \cap U_j$.

Notice that any finite-dimensional real differentiable manifold M satisfies the above conditions. Then,

Proposition 5.10 $\mathbf{L}^B \models \widehat{\tau} \neq \tau^B$.

PROOF. See [21].

Since we are interested in modelling the laboratory situation, we may further restrict our attention to the case when $X = M$, a real, finite-dimensional differentiable manifold; $\mathcal{T}(M)$ notes one of its triangulations. Notice that any triangulation can be always coded by a finite or countably infinite sequence of letters from a discrete alphabet. Therefore, once one such sequence is given in **L**, we can talk about standard triangulations in its Boolean extensions [21].

In what follows we furthermore suppose that M is smoothly embedded into an Euclidean space R^r of sufficiently high dimension. If that is the case we can state:

Proposition 5.11

1. $\mathbf{L}^B \models$ 'Every open ball $U \subset \mathrm{R}^r$ is diffeomorphic to an open ball $W \subset \mathrm{R}^r$ with a constructible center and a constructible diameter'.
2. $\mathbf{L}^B \models$ 'If $K \subset \mathrm{R}^r$ is a compact, smooth submanifold of R^r with or without a boundary, then there is a compact and smooth submanifold $K' \subset \mathrm{R}^r$ with a standard triangulation $(\mathcal{T}(K'))\check{}$ so that K is diffeomorphic to K''.

PROOF. Immediate; see [21].

Spacetime physics is formulated modulo diffeomorphisms. Therefore we have proved that we gain no new objects when we consider open balls and compact domains in our Boolean extension $\mathbf{L}^B$; that result is easily generalized to extensions $\mathbf{L}^C$, where C is any complete Boolean algebra. However we can assert:

Proposition 5.12 $\mathbf{L}^B \models$ 'If $n \geq 2$, there are $N \subset M$, where M is a noncompact differentiable real n-dimensional manifold with a standard triangulation $(\mathcal{T}(M))\check{}$, and N is an open submanifold, such that N cannot be diffeomorphically mapped onto any submanifold with a standard triangulation'.

PROOF. [21].

Yet we also show [21] that:

Proposition 5.13 $\mathbf{L}^B \models$ 'For $n > 1$, if P is R^n without a countable number of nonoverlapping closed balls [that is, R^n minus a countable number of 'holes'], then P is diffeomorphic to an open submanifold of R^n with a standard triangulation'.

The situation here is so rich that the standard flat spacetime R^4 is homeomorphic (but obviously not diffeomorphic) to a set-theoretically generic exotic 4-manifold! The physical meaning of that situation remains unclear [21]. However we can consider the following idealized setting where the laboratory domain may be set-theoretically generic: we first notice that a denumerable number of 'holes' in R^3 can be physically interpreted as R^3 with a denumerable number of particles. Similarly more complicated connectivity properties (handles and the like, which may imply the set-theoretic genericity of the domain) can be associated to Wheeler's 'wormholes' or to string-like 'defects' in the medium. Therefore, if Q is an open, noncompact, idealized domain within a laboratory and if it has a countable infinity of such objects (wormholes or string-like deffects), then we may expect Q might assume a set-theoretically generic (that is, non-Gödel constructive) topology. Such a Q might then be taken as the union of a family of domains where we can define initial values for solutions of physical field equations. The initial values (which would be patched up out of local values in those domains) might again be set-theoretically generic, and would then generate set-theoretically generic fields through the evolution equations.

(The infinity requirement is an unpleasant idealized condition which may be seen as a hindrance in applying those results to the real world. Yet it is present in most treatments of statistical mechanics and quantum field theory, where that condition doesn't lead to difficulties in the step that goes from the mathematical model to the lab situation.)

Now how can we verify that Q is set-theoretically generic? We have suggested [21] [25] that set-theoretic genericity might appear from the physical point of view under the guise of an 'untamable' randomness in Q. However see below another interpretation, where one associates strange attractors to fragments of a Boolean model.

A Bernouilli shift becomes a zero-entropy shift

Ergodic shifts appear in several physical situations. The classical example is Ornstein's proof that geodesic flows on surfaces with negative curvature are Bernouilli flows [36]. A more recent one [4] shows that the Poincaré return mapping for the 'mixmaster' cosmological model is a positive-entropy shift. We give here a simple example of a Bernouilli shift that becomes a zero-entropy shift (and even an object without a definable entropy) when one changes the set-theoretical model.

Again our ground model is Gödel's constructible universe **L**. Within that model, if $p \in \omega_0$, we define shift space to be the product space p^{Z} of all total maps from Z to p. We say that a map $\sigma \in p^{\mathsf{Z}}$ is deterministic if there is an algorithm that generates it. As there is only a countable infinity of such sequences which are nevertheless dense in p^{Z}, when we take them out of the product space we get the space p-Ir of all p-irrational sequences in p^{Z}. As it is well-known, there is an homeomorphism between p-Ir and the space Ir $\subset [0, 1]$ of all irrationals in the unit interval, where both p-Ir and Ir have the corresponding induced topologies.

If we endow p with the equiprobable measure $(1/p, \ldots, 1/p)$ and extend it via the product measure to the whole of p^{Z}, then the Kolmogorov–Sinai entropy per sign of the shift whose phase space is p^{Z} is $h = \log p$. That value remains unchanged when we take out a countable number of paths from p^{Z}, so that the (induced) entropy of p-Ir is again $h = \log p$. We leave to the reader the construction of the corresponding Suppes predicate.

Now let **M** be a Boolean-valued model that satisfies the axiom system in Proposition 5.2; for the actual construction of one such model see [6] [34]. If 2 is the trivial 2-element complete Boolean algebra, we have the identification $\mathbf{L} = \mathbf{L}^2$ and the standard embedding $\mathbf{L} \subset \mathbf{M}$, given by $x \in \mathbf{L} \mapsto \hat{x} \in \mathbf{M}$. Then,

Proposition 5.14 $\mathbf{M} \models$ '$\widehat{p - \mathrm{Ir}}$ is meager and of zero measure in p-Ir in the induced topology'. Therefore, $\mathbf{M} \models$ 'The Kolmogorov–Sinai entropy of the standard portion of the shift vanishes, that is, $h(\widehat{p - \mathrm{Ir}}) = 0$'.

PROOF. From Proposition 5.2.

The situation may be even stranger: if we consider the first model in Proposition 5.3, and if $\mathbf{N}$ is that model, then:

Proposition 5.15 $\mathbf{N} \models$ '$\widehat{p - \mathrm{Ir}}$ is non-Lebesgue measurable in p-Ir; therefore its entropy h cannot be defined'.

PROOF. From 5.3.

Where does randomness come from?

Remark 5.16 We must emphasize several conceptual difficulties that appear when one considers the standard image of the shift $\widehat{p - \mathrm{Ir}}$ in respect to the full shift p-Ir within the expanded set-theoretical universes:

- Certainly the standard image is no Bernouilli shift, as that image has either a zero entropy or isn't even a measurable set.

- However it contains *all trajectories that added up to the fully developed random shift in the original model*, in our case **L**.

  The shift was random in the original model due precisely to those trajectories, *which can be safely excluded from the 'new' shift in the enlarged model without disturbing its Kolmogorov–Sinai entropy*, that is, without destroying its randomness. Randomness therefore appears to result solely from the existence of generic trajectories, which didn't appear in the constructive universe.

- Therefore we can ask: where does randomness comes from in a Bernouilli shift? Certainly not from its trajectories, when seen in isolation or altogether. It comes from the *relation* among those trajectories themselves; it is a relational property, not something that arises out of individual trajectories.

Randomness, and Bernouilli-shift randomness, intuitively arise out of, say, an (idealized, two-sided infinite) game of heads and tails. Therefore it *should* somehow 'belong' to the actual trajectories in the shift. But, as we have shown, that certainly isn't the case.

Therefore, what is randomness?

Set-theoretically generic electromagnetic fields

The following treatment is slightly different from the one given in [11], but the results are the same. Let M be a spacetime manifold, that is, a 4-dimensional Hausdorff real differentiable manifold endowed with a Lorentzian metric (a real symmetric nondegenerate bilinear form of signature $(- + ++)$. For 'differentiable' we mean here 'of class C^∞'. We do not require the metric for the definition of an electromagnetic field, but we are going to need it when we write down the Maxwell field equations.

We restrict our attention to an open ball $U \subset M$. We are also going to need the function space $C^\infty(U, \mathrm{C})$ of all complex-valued C^∞ functions on U; we endow that space with the usual Fréchet topology [46]. Thus $C^\infty(U, \mathrm{C})$ becomes a Polish space, and we can apply all our previous results on Polish spaces and their category properties to it. Since $C^\infty(U, \mathrm{C})$ is our main function space, from here on we are going to abbreviate it by $\mathcal{F}$.

Let $\bigwedge^2(U, \mathrm{C})$ be the bundle of all C-valued 2-forms on U; its space of smooth sections is $C^\infty \bigwedge^2(U, \mathrm{C}) = \mathcal{F} \otimes \bigwedge^2$, where $\bigwedge^2$ is the vectorspace of bivectors in the tensor product $\mathrm{R}^4 \otimes \mathrm{R}^4$. The topology on $\mathcal{F} \otimes \bigwedge^2$ is the co-induced topology via the tensor product; it is also a Polish topology.

Definition 5.17 $\phi \in \mathcal{F} \otimes \bigwedge^2$ is an **electromagnetic field** if and only if $d\phi = 0$, where d is the exterior derivative operator.

See on that definition [29], p. 138.

The corresponding Suppes predicate is:

$$E(X) \leftrightarrow \exists\phi\exists U(X = \langle\phi, U\rangle \wedge \phi \in \mathcal{F} \otimes \bigwedge{}^2 \wedge\, d\phi = 0),$$

where one supposes given the Suppes predicates for U, $\bigwedge^2$, the function space $\mathcal{F}$ and related objects.

One reads '$E(\langle\phi, U\rangle)$' as 'ϕ is an electromagnetic field over U'.

Remark 5.18 Notice that $\mathbf{L}^B \models E(\langle\phi, U\rangle)$ if and only if $\|E(\langle\phi, U\rangle)\| = 1$, again if and only if both

$$\|\phi \in \mathcal{F} \otimes \bigwedge{}^2\| = 1$$

and

$$\|d\phi = 0\| = 1.$$

Also, Maxwell's equations can be written:

$$\begin{cases} \delta\phi = \iota \\ d\phi = 0 \end{cases}$$

where ι is the smooth complex-valued 1-form that represents the field's 'current' or 'source', and δ is the (dual) divergence with respect to d. Concerning that form of the Maxwell system see [26]. (It is worth noting that most classical field equations in physics can be written in such a Maxwellian or Dirac-like form.)

As usual, $\delta = *d*$, $*$ being Hodge's star operator on M ($*$ depends on the metric for M). The Maxwell equations can also be written as:

$$\nabla\phi = \iota,$$

with $\nabla = d - \delta$. ∇ is the Dirac operator on M, and we suppose the exterior bundle endowed with a Clifford product [26][28].

Denote by ker d the kernel of the exterior derivative that acts on M's exterior bundle. Then $E_U = \ker d \subset \mathcal{F} \otimes \bigwedge^2$ is the space of all electromagnetic fields on U; it is an obvious ZFC set, and one has:

Proposition 5.19 E_U is Polish in the induced topology.

PROOF. $E_U = d^{-1}(0)$ is closed in $\mathcal{F} \otimes \bigwedge^2$. It is therefore Polish in the induced topology [3].

Now let $\mathbf{L}^C$ be as in Corollary 5.8, and $\mathbf{L}^B$ and $\mathbf{L}^{B'}$ as in Corollary 5.9. We have that:

Proposition 5.20 $\mathbf{L}^C \models$ 'The set of nonconstructive (or set-theoretically generic) electromagnetic fields on U is topologically generic, that means, it has as a complement a meager set'.

PROOF. From Corollary 5.8.

Similarly, we have:

Proposition 5.21

1. $\mathbf{L}^B \models$ 'If φ denotes the Polish space projection $\varphi : \omega_0^{\omega_0} \to E_U$, and if $\widehat{E_U}$ are the constructible electromagnetic fields, then both $\varphi^{-1}(\widehat{E_U})$ and its complement in the covering space aren't measurable'.
2. $\mathbf{L}^{B'} \models$ 'Both the sets of constructive and nonconstructive electromagnetic fields on U do not have the Baire property'.

PROOF. From Corollary 5.9. ∎

Set-theoretic genericity in physics

Remark 5.22 The preceding results highlight the following aspects of set-theoretic genericity, when seen from the point of view of a physical theory:

- Set-theoretic genericity does not depend on nonlinear structures, as the Maxwell equations form a linear system. That fact reminds one of the folklore conjecture that relates undecidability, incompleteness, and nonlinearity: another recent work of us [12] again shows that those properties are unrelated.

- Also, set-theoretically generic fields do not necessarily arise out of generic sources; it suffices, say, to consider all vacuum solutions for the electromagnetic field equations in a Boolean extension $\mathbf{L}^D$, where D is a c.c.c. complete Boolean algebra. Here we will have set-theoretically generic solutions out of a trivially constructive object, the zero or vacuum source. The extension to nonhomogeneous systems is immediate.

- The fields that we introduce when we enlarge our model are *physically* different, once we suppose that different closed 2-forms are different electromagnetic fields.

- They cannot in general be reached by the usual approximation procedures out of a smaller, previously known set, since approximation procedures are essentially constructive procedures [24].

- Finally, the set-theoretically generic fields will also be topologically generic in the model $\mathbf{L}^C$, but won't even be a Baire space in the induced topology in $\mathbf{L}^{B'}$. Therefore set-theoretic genericity and topological genericity do not necessarily match, when one deals with Polish spaces.

Note added in May 1995. The authors have recently explored the relation between forcing models and metamathematical phenomena in the axiomatized sciences; see [19] [20].

An interpretation: strange attractors?

We are now going to sketch a few ideas about a possible geometric interpretation for some Boolean extensions. We wish to address the following question: how do we picture, in a more concrete (or perhaps even physically realizable) way, a Boolean-valued set in a Boolean extension? What kind of reality—besides *formal* reality—does a Boolean-valued physical field possess?

Consider Figure 1 below. It represents the local structure of a 'strange attractor' for a dynamical system (say, the local structure of Hénon's map [30] p. 269); the vertical axis has a Cantor set C while the horizontal axis is an interval I; locally, the strange attractor is $C \times I$.

A Cantor set has cardinality $2^{\aleph_0}$; it can be seen as the Stone space for the Boolean algebra $B = \mathrm{RO}(2^{\aleph_0})$, which is isomorphic to the clopen sets in the Cantor set.

Thus the algebra's 1 is the whole Cantor set.

Figure 2 translates the following Boolean truth value: if x denotes the point on the horizontal axis, then in order to represent $\|x \in I\| = 1$ one has to take the whole 'fiber' over x. Again one knows that in order to have $\|u \subset I\| = 1$ it is enough that $0 \leq \|x \in u\| \leq 1$, for all $\|x \in I\| = 1$.

That condition is represented in Figure 3, where the Boolean-valued subset appears as a set of chaotic trajectories in the attractor; the smaller the Boolean value

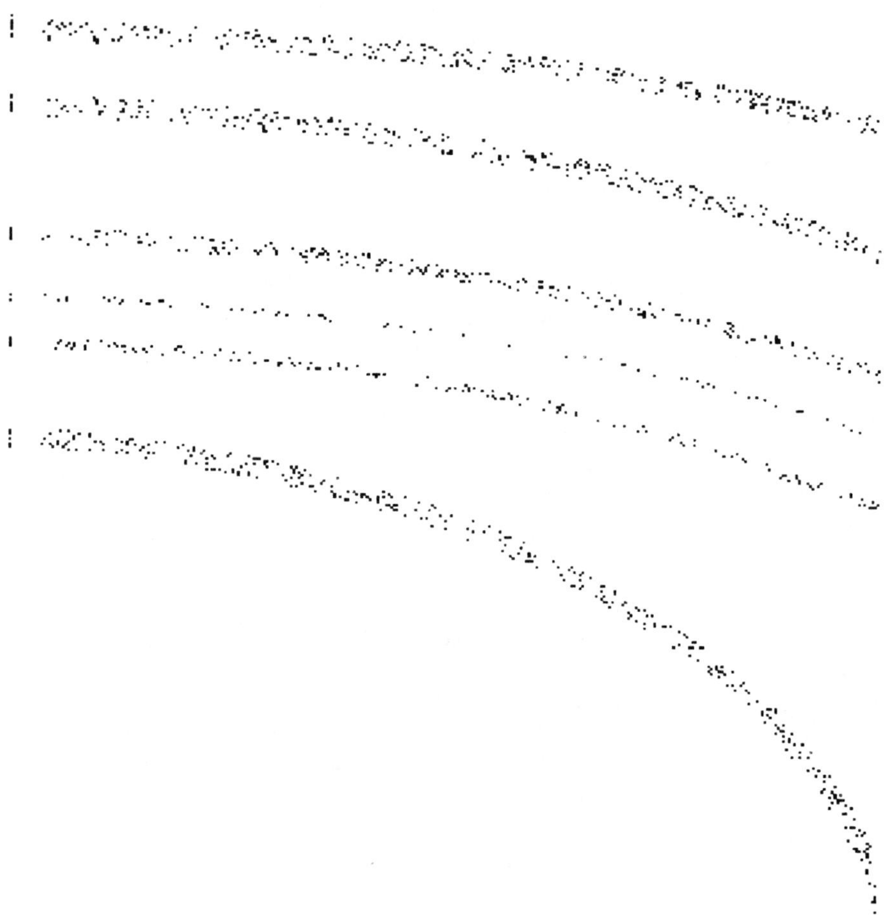

Figure 1: *Locally a strange attractor looks like the product $C \times I$ of a Cantor set and an interval.*

Figure 2: *If x is the point on the horizontal axis, then the Boolean value $\|x \in I\| =$ 1 is given by the whole fiber over x, which reproduces the Cantor set that represents 1 in the algebra.*

$b = \|x \in u\|$, the more precisely one determines the chaotic trajectory or set of trajectories

However one must notice that since Boolean values are represented by *clopen* subsets of the Cantor set, it would be interesting to see whether there is a 1-1 relation between trajectories in the attractor and representations like the one in Figure 3; at least the cardinality of the set of trajectories is $2^{\aleph_0}$, which suggests a possible match. If they don't match, one could also consider approximation procedures and the like.

That kind of representation is very convenient, and has the advantage of suggesting a simple visual connection between (a fragment) of a Boolean-valued model, and chaos theory in dynamical systems. We are fully conscious of the fact that most Boolean models arise out of Boolean algebras of very high cardinality, and we certainly wouldn't know how to incorporate those huge algebras into the present picture. Also, we have recently answered in the negative a question by M. Hirsch that asks for (algorithmic) criteria to check whether a given dynamical system is chaotic or not; our answer shows that there are intractable innocent-looking systems such that it is either arbitrarily difficult to determine any of its properties, or even impossible, due to Gödel incompleteness, to prove within any reasonable axiom system those properties [12] [32]. Therefore an actual implementation of that picture of ours might be extremely difficult.

Nevertheless its *Gestalt*-like pregnancy seems to impose itself, and we excuse ourselves for presenting it here in such an incomplete way.

Developments since 1991

Note added in May 1995. After the conclusion of the original version of this paper in late 1990, the authors obtained a series of results on the undecidability and incompleteness of physical theories. They are summarized below; for a detailed presentation see [19] and the references quoted.

The integrability problem in classical mechanics

That's an old question. We quote a recent reference on it (already mentioned in[11]):

> Are there any general methods to test for the integrability of a given Hamiltonian? The answer, for the moment, is no. We can turn the question around, however, and ask if methods can be found to construct potentials that give rise to integrable Hamiltonians. The answer here is that a method exists, at least for a restricted class of problems. . .

We can divide the integrability question into three queries:

- Given any Hamiltonian h, is there an algorithm to decide whether the associated flow X_h can be integrated by quadratures?

Figure 3: *A Boolean-valued subset as a set of chaotic paths through the attractor.*

- Given an arbitrary Hamiltonian h such that X_h can be integrated by quadratures, can we algorithmically find a canonical transformation that will do the trick?

- Can we algorithmically check whether an arbitrary set of functions is a set of first integrals for a Hamiltonian system?

No, in all three cases. There is no general algorithm to decide, for a given Hamiltonian, whether or not it is integrable. Also there will be sentences such as $\xi =$ 'h is integrable by quadratures', where however $T \nvdash \xi$ and $T \nvdash \neg\xi$ [11].

The Hirsch problem: the decision problem for chaos

That problem was discussed in the introduction. Is there an algorithm to check for chaos given the expressions of a dynamical system? No: there is no such a general algorithm, and there will be systems that look chaotic on a computer screen (that is to say, they are chaotic in a standard model **M**, but such that proving or disproving their chaotic behavior is impossible in T [11] [32].

That result applies to *any* nontrivial characterization for chaos in dynamical systems. Let U be an open set in phase space. Then:

Proposition 5.23 There is no general algorithmic procedure to check:

1. Whether an arbitrary vectorfield X over U is ergodic.

2. If $\dim M \geq 4$, whether an arbitrary vectorfield X over U has a Smale horseshoe.

3. If M is compact, real, two-dimensional, of class C^3 and has a constant negative curvature, whether an arbitrary X is a Bernouillian flow.

The corresponding incompleteness results are:

Proposition 5.24 Within ZFC:

1. There is an expression for a motion $m(t)$ on R^2 such that the assertion '$m(t)$ is ergodic in R^2', can neither be proved nor disproved in ZFC.

2. There is an expression for a differentiable dynamical system v such that the statement 'v has a Smale horseshoe', can neither be proved nor disproved from the axioms of ZFC.

3. There is a flow X so that the statement 'X is a Bernouillian flow', can neither be proved nor disproved from the ZFC axioms.

4. If ϕ is a ZFC predicate that gives a nontrivial characterization for chaos in a dynamical system, then there is an expression Z for a dynamical system such that neither $\phi(Z)$ nor $\neg\phi(Z)$ can be proved in the formal theory.

Wolfram's conjecture and Penrose's thesis

Wolfram had long conjectured that simple phenomena in classical physics might lead to undecidable questions [48]:

> One may speculate that undecidability is common in all but the most trivial physical theories. Even simply-formulated problems in theoretical physics may be found to be provably insoluble.

On the other side Penrose asserted as a kind of thesis that classical physics offers no examples of noncomputable phenomena. The preceding results prove Wolfram's conjecture and therefore give a counterexample to Penrose's thesis [12] [41].

Arnol'd's problems

Arnol'd formulated in the 1974 AMS Symposium on the Hilbert Problems [2] a question dealing with algorithmic decision procedures for polynomial dynamical systems over Z:

> *Is the stability problem for stationary points algorithmically decidable?* The well-known Lyapounov theorem solves the problem in the absence of eigenvalues with zero real parts. In more complicated cases, where the stability depends on higher order terms in the Taylor series, there exists no *algebraic* criterion.
>
> *Let a vector field be given by polynomials of a fixed degree, with rational coefficients. Does an algorithm exist, allowing to decide, whether the stationary point is stable?*
>
> A similar problem: Does there exist an algorithm to decide, whether a plane polynomial vector field has a limit cycle?

For those questions there is no general algorithm available if we add the sine function, the absolute value function and π to our language; also the corresponding theory is incomplete [15]. In the polynomial case again there is no algorithm to decide whether a fixed point at the origin is stable or not [16] [17].

'Smooth' problems equivalent to hard number-theoretic problems

Common wisdom among mathematicians has that number-theoretic problems are in general much more difficult than 'smooth' problems. We showed that this is definitely not the case. We gave an explicit example of a dynamical system where the proof that there will be chaos is equivalent to the proof of Fermat's last theorem (or the proof of Riemann's hypothesis). We also proved that (given some conditions) those 'nasty' problems are dense in the space of all dynamical systems [22].

Problems worse than any number-theoretic problem

The language of analysis is much richer than the language of arithmetic, as we can express the halting function in analysis. Also we can explicitly construct 'natural'-looking problems which lie beyond the arithmetical hierarchy with our techniques [15].

Generic, faceless objects

One of the features of the main set-theoretic forcing constructions is that we add 'generic', faceless sets to our formal theories. However there are no explicit expressions for those objects. With the help of our techniques we exhibited an expression for a 'faceless' Hamiltonian [15]: the only thing we can prove about it is that it definitely is a Hamiltonian, and nothing more, as shown above.

The relation between forcing and our techniques is discussed in the previous section; see also [14] [18].

Undecidable sentences as bifurcation points in a formal theory

Our techniques allow a simple coding of undecidable sentences in T into a bifurcating vectorfield [16] [17]. The example sketched in the two previous subsections goes (roughly) as follows: given the extended theories $T^+ = T + P(\xi)$, $T^- = T + \neg P(\xi)$, we have that:

- $T^+ \vdash P$ if and only if for a certain vectorfield X, $T^+ \vdash$ 'X undergoes a Hopf bifurcation';
- $T^- \vdash \neg P$ if and only if $T^- \vdash$ 'X doesn't undergo a Hopf bifurcation'.

Notice that we can algorithmically obtain an expression for that X within our formal language. More about that in [17] [18]. That construction can be fully developed into a functor from formal systems to dynamical systems, where Gödel incompleteness translates into bifurcation phenomena.

Those results are now slowly making their way into the mainstream; see [9] [42].

6 Conclusion

We emphasize the following point: we have only touched 'inessential' or 'unusual' set-theoretic axioms, such as the continuum hypothesis, the axiom of constructibility or Martin's axiom; therefore most of classical mathematics remains untouched by them and is kept in our models. Our more recent research quoted above [11] [12] [13] [23] [21] [24] shows that there are infinitely many 'meaningful' undecidable sentences within any empirical science axiomatized through Suppes predicates, but we have only started the investigation of the physical interpretation of forcing models and the consequences of its undecidable sentences for that empirical science.

7 Acknowledgments

It is certainly a pleasure to dedicate the present paper to Professor V. Smirnov on his 60th birthday.

The authors thank the Institute for Advanced Studies, University of São Paulo, for partial support of the present work. The second author also acknowledges grants from FAPESP, CNPq-Brazil, and the Fulbright Commission (for the period 1989–1990, when this paper was first drafted). FAD also thanks Professor Patrick Suppes for his hospitality at the IMSSS, Stanford University, and for many discussions and criticisms on this and related work; and last but not least the School of Communications in Rio de Janeiro, and CETMC, its laboratory voted to mathematical models in communications, for their support.

Formatting of this paper is due to *Project Griffo* and to the Graduate Studies Program Computer Laboratory, School of Communications, Federal University at Rio de Janeiro.

Federal University of Rio de Janeiro, Brazil

References

[1] S. Albeverio, J. E. Fenstad, R. Høegh-Krohn, T. Lindstrøm. *Nonstandard Methods in Stochastic Analysis and Mathematical Physics*. Academic Press, 1986.

[2] V. I. Arnol'd. Problems of Present Day Mathematics, XVII (Dynamical systems and differential equations). *Proc. Symp. Pure Math.* **28**, 59, 1976.

[3] W. Arveson. *An Invitation to C^* Algebras*. Springer, 1976.

[4] J. D. Barrow. *Phys. Rep.* **85**, 1, 1982.

[5] N. Bourbaki. *Theory of Sets*. Hermann & Addison-Wesley, 1968.

[6] J. D. Bell. *Boolean-Valued Models and Independence Proofs in Set Theory*. Clarendon Press, Oxford, 1984.

[7] P. A. Benioff. *J. Math. Phys.* **32**, 618, 1976.

[8] P. A. Benioff. *J. Math. Phys.* **32**, 629, 1976.

[9] J. Casti. *Reality Rules*, I and II. John Wiley, 1992.

[10] N. C. A. da Costa and R. Chuaqui. *Erkenntnis* **29**, 95, 1988.

[11] N. C. A. da Costa and F. A. Doria. Suppes Predicates for Classical Physics. In J. Echeverría et al., eds., *The Space of Mathematics*. de Gruyter, 1992.

[12] N. C. A. da Costa and F. A. Doria. *Int. J. Theor. Phys.* **30**, 1041, 1991.

[13] N. C. A. da Costa and F. A. Doria. *Found. Phys. Letters*, **4**, 363, 1991.

[14] N. C. A. da Costa and F. A. Doria. *Philosophica*, **50**, 901, 1992.

[15] N. C. A. da Costa and F. A. Doria. Suppes Predicates and the Construction of Unsolvable Problems in the Axiomatized Sciences. In P. Humphreys, ed., *Patrick Suppes, Scientific Philosopher*, II, Kluwer, 1994.

[16] N. C. A. da Costa and F. A. Doria. On Arnol'd's Hilbert Symposium Problems. In G. Gottlob, A. Leitsch, D. Mundici, eds., *Proceedings of the 1993 Kurt Gödel Colloquium: Computational Logic and Proof Theory*, Lecture Notes in Computer Science **713**. Springer, 1993.

[17] N. C. A. da Costa and F. A. Doria. *Int. J. Theor. Phys.*, **33**, 1913, 1994.

[18] N. C. A. da Costa and F. A. Doria. Gödel incompleteness in analysis, with an application to the forecasting problem in the social sciences. *Philosophia Naturalis*, **31**, 1, 1994.

[19] N. C. A. da Costa and F. A. Doria. Gödel Incompleteness, Explicit Expressions for Complete Arithmetic Degrees and Applications. To appear in *Complexity*, 1995.

[20] N. C. A. da Costa and F. A. Doria. From the Halting Function to Set-Theoretic Forcing. Preprint, 1995.

[21] N. C. A. da Costa, F. A. Doria and J. A. de Barros. *Int. J. Theor. Phys.*, **29**, 935, 1990.

[22] N. C. A. da Costa, F. A. Doria and A. F. Furtado do Amaral. *Int. J. Theor. Phys.*, **32**, 2187, 1993.

[23] N. C. A. da Costa, F. A. Doria, A. F. Furtado do Amaral and J. A. de Barros. *Found. of Phys.*, **24**, 783, 1994.

[24] N. C. A. da Costa, F. A. Doria and J. A. de Barros. A Formally Undecidable Statement in Classical Electromagnetic Theory. In C. A. Bertulani and J. Lopes Neto, eds., *Encontro de Física Teórica do Rio de Janeiro–Homenagem Póstuma ao Prof. Carlos Marcio do Amaral*. UFRJ, 1990.

[25] N. C. A. da Costa, F. A. Doria and J. A. de Barros. 'Suppes Randomness in Arbitrary Spaces. Preprint, preliminary version, 1990.

[26] F. A. Doria. *J. Math. Phys.*, **18**, 564, 1977.

[27] F. A. Doria. A Bifurcation Set Associated to the Copy Phenomenon in the Space of Gauge Fields. In G. I. Zapata, ed., *Functional Analysis, Holomorphy and Approximation Theory II*. North-Holland, 1984.

[28] F. A. Doria, S. M. Abrahão and A. F. F. do Amaral. *Progr. Theor. Phys.*, **75**, 1440, 1986.

[29] G. G. Emch. *Mathematical and Conceptual Foundations of 20th Century Physics*. North-Holland, 1984.

[30] J. Guckenheimer and P. Holmes. *Nonlinear Oscillations, Dynamical Systems and Bifurcations of Vector Fields*. Springer, 1983.

[31] D. Hilbert. Mathematical Problems. In F. E. Browder, ed., *Mathematical Developments Arising from the Hilbert Problems*. Proc. Symp. Pure Math. **28**, AMS, 1976.

[32] M. W. Hirsch. The Chaos of Dynamical Systems. In P. Fischer and W. R. Smith, editors, *Chaos, Fractals and Dynamics*. Marcel Dekker, 1985.

[33] T. Jech. *Set Theory*. Academic Press, 1978.

[34] K. Kunen. *Set Theory*. North-Holland, 1983.

[35] C. A. Lungarzo. Superposition of States in Quantum Logic from a Set-Theoretical Point of View. In A. I. Arruda, N. C. A. da Costa and R. Chuaqui, eds., *Mathematical Logic—Proceedings of the First Brazilian Conference*, Lecture Notes in Pure and Applied Math. Marcel Dekker, 1976.

[36] D. Ornstein and B. Weiss. *Israel J. Math.*, **14**, 185, 1973.

[37] J. C. Oxtoby. *Measure and Category*. Springer, 1980.

[38] M. B. Pour-El and I. Richards. *Adv. Math.*, **39**, 215, 1981.

[39] J. R. Shoenfield. *Amer. Math. Monthly*, **82**, 610, 1975.

[40] R. M. Solovay. *Ann. Math.*, **92**, 1, 1970.

[41] I. Stewart. *Nature*, **352**, 664, 1991.

[42] I. Stewart. *The Problems of Mathematics*, 2nd. ed. Oxford, 1992.

[43] P. Suppes. *Introduction to Logic*. Van Nostrand, 1957.

[44] P. Suppes. *Set-Theoretical Structures in Science*, mimeo. Stanford University, 1967.

[45] P. Suppes, *Scientific Structures and their Representation*, preliminary version. Stanford University, 1988.

[46] F. Trèves. *Topological Vector Spaces, Distributions and Kernels*. Academic Press, 1967.

[47] A. S. Wightman. Hilbert's Sixth Problem: Mathematical Treatment of the Axioms of Physics. In F. E. Browder, ed., *Mathematical Developments Arising from the Hilbert Problems*, Proc. Symp. Pure Math. **28**. AMS, 1976.

[48] S. Wolfram. *Commun. Math. Phys.*, **96**, 15, 1984.

[49] J. D. Maitland Wright. *Bull. AMS*, **79**, 1247, 1973.

Part III

Logical Semantics

8

World Lines and their Role in Epistemic Logic

Jaakko Hintikka

1 Introduction

This paper is an excerpt from a longer work which so far remains incomplete and unpublished. In view of the intrinsic interest of the subject matter, I believe it is motivated to publish it independently.

The notation used is fairly standard. As usual, 'knows that' has as its formal counterpart 'K'. The main difference as compared with my earlier publications is that the known indicated by a bracketed singular term in front of 'K', not by a subscript. Thus

$$a \text{ knows that } S \tag{1}$$

will be expressed by

$$(a)KS \tag{2}$$

not by '$K_a S$.

The usual model sets and model systems approach is used, with the usual conditions and rules for propositional epistemic logic. This approach can easily be converted into a Gentzen-style treatment of epistemic logic. However, that conversion is not discussed here.

There is one difference, however, in the techniques used to construct model sets. With each approximation μ to a model set, we now associate another set $\bar{\mu}$ which comprises those formulas of which we already know at the current stage of the construction that they must not occur in the model set to be constructed. Thus the technique used here represents a kind of synthesis of the model set techniques and Beth's *tableau* method. Indeed, μ corresponds to the left column of a Beth *tableau* and $\bar{\mu}$ to its right column.

Rules are easily formulated for the new combined construction technique. Indeed, the rules for $\bar{\mu}$ are simply mirror images (duals) of the rules for μ. The rules dealing with the approximation to a model set are designated $(C.-)$, those dealing with what is being excluded from the approximation $(\bar{C}.-)$.

P. I. Bystrov and V. N. Sadovsky (eds.), Philosophical Logic and Logical Philosophy, 121–137.

The model described by the model set μ is referred to as $||\mu||$. The negation fo the membership relation $\in$ is expressed by $\bar{\in}$.

If a symbol occurs in a formula outside the scopes of all intensional operators, which in this work means outside the scopes of all the K- operators, it is said to occur in that formula *extensionally*.

2 The role of world lines in *knowing who* or *what*

Careful explanations are needed in order to do justice to all different types of knowledge. The usual basic logic of knowledge shows us how to handle *knowing that*. How can we deal with *knowing who* and other similar *knowing* + *wh*-constructions? The first idea is clear. To say that x knows who, say y, satisfies the condition $S[y]$, is clearly to say that x knows of some particular individual z that $S[z]$. In other words, the structure of *knows* + *wh*-construction sentences is

$$(\exists z)\ (x)K\ S[z]. \tag{3}$$

In a rough-and-ready way, we can thus say as a slogan that on our analysis the logic of the *knows* + *wh*-constructions is the logic of quantifying into epistemic contacts. Certain ramifications of this basic idea have been registered in the literature. They are due to the two-barrelled character of wh-worlds in natural languages. Later in this paper, we shall also find another important way in which the same basic idea has to be carried further.

What (3) amounts to in the first place is an analysis of *knows* + *wh*-constructions in terms of *knowing that*. This analysis has to be taken with a grain of salt, however.

While (3) brings out a genuine connection between *knowing who* (*what, where, when*, etc) and *knowing that*, more is involved in the model theory of sentences of the form (3) than in most *knowing that* sentences. This additional ingredient in our semantics for epistemic notions is important, and it makes us hesitant to speak of an outright reduction of *knowing who* to *knowing that*.

In order to see what this new ingredient is, let us note first that (3) we are clearly speaking of *one and the same individual* z as a member of all the different epistemic x-alternatives which the operator $(x)K$ evokes. This means that there must be more structure in an epistemic space than we have specified so far. Identities of individuals in different models in the space must be determined. We can envisage of this determination taking the form of a warp of world lines connecting the different models in an epistemic space.

Now it is possible to see why we need two different kinds of singular terms. We need terms which pick out the same individual from all those worlds from which they pick out anything at all, and terms which may designate different individuals in different worlds. Constants proper serve the former purpose, ordinary (improper) constants the latter.

Our proper constants are obviously related closely to Russell's 'logically proper names' and to Kripke's 'rigid designators'. Both those earlier notions carry misleading associations, however, and in order to avoid them we have chosen to employ terms different from Russell's and Kripke's. (Cf. below, this paper, sections 3–4.)

Problems concerning logically proper names, rigid designators, world lines etc. have prompted a fair amount of discussion. The line of thought which has led us to these problems nevertheless leads to a further conclusion which has not received the attention in the literature which it deserves.

3 World lines can break down

What the apparatus of world lines of cross-identification enables us to do is among other things to make a distinction between those constants b of whose bearers x knows who (what, where, etc.) they are, and those of which x does not know who (what, where, ...) they are. In the former case, the referent of 'b' coincides with the embodiment of one and the same individual in each of x's knowledge worlds, in other words, the referent of 'b' follows one and the same world line in all these worlds.

This distinction is not the only one we have to make, however. One can know or fail to know, not only who or what the bearer of an improper constant is, but also know or fail to know who or what some one individual (bearer of a proper constant) is. For this distinction, there is no room in the existing treatments of epistemic logic. Yet it is a very real phenomenon. Let's first consider an example, or more accurately speaking, a small scale analogue of the breakdown problem. Let us consider, not possible worlds, but possible (and indeed, frequently actual) basketball games. They are now our counterparts to 'possible worlds'. Then the players on the court will correspond to the denizens of the different worlds (models) one is considering in an epistemic situation. Having seen a number of basketball games, in a very reasonable sense I know who Julius Erving is. There is no epistemically possible basketball game, as far as I am concerned, where I could not locate Dr. J on the floor, if he is there. However, I do not in the same sense know who some of the less unique basketball players are. Given the name of such a player and a game I am watching, I don't know how to go about ascertaining whether he is one of the players on the floor. Yet I don't believe that some particular player unknown to me could be one person in one of my epistemic worlds and another person in another world. How can my epistemic situation be represented model-theoretically?

Notice that this problem has nothing to do with the possible failure of existence. An individual's failure to exist in some particular possible world is the analogue of some particular player's not getting playing time in some one game. Clearly this has nothing to do with my success or failure to know who the players are who are on some particular occasion—an actual occasion or a possible occasion compatible with everything I know—down there on the floor playing.

It nevertheless is not hard to see what the general form of the puzzle is. We have tried to explain *knowing who b is* by means of b's following a world line. If this is what we have to do, how can we deal with the situation in which I know that a term follows a well-defined world line, but in which its bearer's identity is nevertheless unknown to one?

A solution is in fact quite easy, once the nature of the problem is realised. I can be blissfully ignorant of the identity of an individual that always goes together with a world line, provided that world lines can break down in moving from one of my epistemic alternations to another. Nothing could be simpler than this assumption. The observations made in the last few paragraphs imply, however, that the breakdown of a world line when one moves to a world w does not mean merely that the individual x whose identity is defined by this world line fails to exist in w. It means something much more radical. It means that it does not even make sense to ask whether x exists or does not exist in w.

This is nevertheless perfectly in keeping with our example. My failure to know who some of the members of a basketball team are does not mean merely that I cannot recognise them when they are in a game. It is also reflected by the fact that I cannot tell whether they are playing in some particular game or not. Moreover, the two failures clearly are but two sides of the same coin. It is hard to imagine how I could be able to carry out one kind of recognition but the other.

4 Examples

The need of allowing world lines to break down in *this* sense is also shown by more technical examples. If we don't modify the usual conditions on model sets and model systems, the following will be a truth of epistemic logic:

$$(\exists x)K(y = x) \tag{4}$$

(We are omitting here the indications of the knower.) But what (4) says is that everybody knows who everybody is, which is nonsense.

The alleged (unreconstructed) 'logical truth' of (4) can be shown as follows:

1. $(\exists x)K(y = x) \in \bar{\rho}, \rho \in \Omega$ (Counter-assumption)
2. $K(y = y) \in \bar{\rho}$ (from 1. by (C. E))
3. $(y = y) \in \bar{\lambda}$ (λ is an epistemic alternative to ρ; from 2. by $(\bar{C}.K)$)
4. $(y = y) \in \lambda$ (from 3. by (C.self =))

Here 3. and 4. violate (C. comp) which in effect says that one and the same sentence cannot be both true and false, completing the proof.

Notice that the only reasonable-looking way of avoiding this proof is to disallow $(y = y) \in \lambda$. We shall argue later that this is indeed the reasonable way of treating failures of world lines (of the kind we are dealing with here).

Other similar examples are not hard to find. Here is one:

$$(\forall y)(\exists u)(y = u\&(\exists z)(a)K(u = z)) \tag{5}$$

It can be shown to be logically true by means of the conditions (and the corresponding rules for constructing model sets and model systems), as the reader can easily ascertain, unless changes are made. Yet what (4) clearly says is that a knows who each individual is, which clearly is not a logical truth, not even a material truth in most applications. The logical truth of (5) can be shown as follows:

1. $(\forall y)(\exists u)(y = u\&(\exists z)K(u = z)) \in \bar{\rho}, \rho \in \Omega$ (counter-assumption)
2. $(\exists u)(v = u\&(\exists z)K(u = z)) \in \bar{\rho}$ (from 1. and by $(\bar{C}.U)$)
3. $(\exists u)(v = u) \in \rho$ (from 1. by $(\bar{C}.U)$)
4. $(v = v)\&(\exists z)K(v = z) \in \bar{\rho}$ (from 2., 3. by $(C.E)$)

First alternative line of construction:

5. $(v = v) \in \bar{\rho}$ (from 4. by $(\bar{C}.\&)$)
6. $(v = v) \in \rho$ (from 3. by $(C.\text{self} =)$)

Here (5) and (6) contradict $(C.\text{comp})$ closing this line of argument.

7. $(\exists z)K(v = z) \in \bar{\rho}$ (from 4. by $(\bar{C}.\&)$)
8. $K(v = v) \in \bar{\rho}$ (from 7. by $(\bar{C}.E)$)
9. $(v = v) \in \bar{\lambda}$ ($\lambda \in \Omega$ is an epistemic alternative to ρ) from 8. by $(C.K)$)
10. $\sim (v = v) \in \bar{\lambda}$ (from 3. by $(C.\text{self} =)$)

Here 9. and 10. contradict $(C.A \sim)$, completing the argument. Indications of the knower have again been omitted from the argument.

5 Describing the breakdown of world lines

But how is the breakdown of a world line, say the world line associated with a proper constant 'x', to be handled in the model set technique? Or, to put the same question in a different way, how is the fact that this world line can be continued to $\|\rho\|$ to be expressed? If the world line can be so continued, we shall say that x is *well defined* in $\|\rho\|$. How can this well definedness be 'seen' from ρ?

In an earlier paper, Jaakko Hintikka tried an approach in which the mere presence of 'x' in the members of ρ extensionally expresses the well-definedness of x in $\|\rho\|$. However, this approach leads to considerable complications in the quantifier rules. It is therefore advisable to try a simpler treatment. The idea adopted here is that the well- definedness of x in $||\rho||$ is to be expressed by $(x = x) \in \rho$. This creates a kind of parity between the two kinds of failures of world line. The world line of x fails to continue to $||\rho||$ because of a failure of well-definedness if $(x = x)\bar{\varepsilon}\rho$. It fails because of a failure of existence if $(\exists y)(x = y)\bar{\varepsilon}\rho$ and $(\exists y)(y = x)\bar{\varepsilon}\rho$ for all 'y'. The difference between two kinds of failure to be there in a world is thus essentially the same as the difference between $=$ and $\exists$.

The observations made above allow even a kind of quasi-deduction of our formal treatment of the breakdown of world lines. Being able to decide in principle whether y exists in a given world w_1 clearly means being able to compare y in turn with each individual in the domain of w_1 for identity. If one of them is identical with y, y exists in w_1. Otherwise it does not exist there. Thus being able to decide whether y exists in a world w_1 is the same as to be able to compare y for identity with the denizens of w_1. But if you can compare y with them, surely you can compare y with itself. If you can, surely it will be the case that $y = y$ in w_1. If you cannot, you cannot assert that $y = y$ in w_1. Hence the natural expression for the well-definedness of y in $||\rho||$ is $(y = y) \in \rho$. (Notice that this does not commit one to the existence of y in $||\rho||$.)

One way of formulating what we are doing is to distinguish knowing the identity of someone, say b, *de dicto* and *de re*. The former means knowing who 'b' refers to. The latter means knowing of the individual who in fact is b who he or she is. This contrast is obviously related closely also to the contrast between knowing who the bearer of an improper constant is and knowing who the bearer of a proper constant is. This contrast makes in fact good interpretational sense. For instance, in some sense a proper name picks out the same individual in all worlds. However, people frequently fail to know who the bearer of a self- acknowledged proper name is. How is such ignorance to be expressed? Our treatment provides an answer. It amounts to an important new stage in clarifying the meaning of such locutions as *knowing who* someone is.

In earlier books and papers, Jaakko Hintikka has dealt with the problem of knowing who *de dicto*, we might say. What we are doing here is to give an analysis of *knowing* + *wh*-construction in its *de re* sense.

Here we can also see what's wrong with Kripke's use of his idea of rigid designator. Even if a designator can never be 'bent', it can fail to be operational in some world or other. The reason why Kripke overlooks this possibility is that he does not take seriously the problem of drawing world lines. This refusal is merely a philosophical prejudice, however. The idea that the identification of individuals between different worlds is somehow given to us by the grace of logic or God or some other authority is an example of the worst kind of arbitrary metaphysics. But if one tries to give a realistic account of cross-identification, one soon sees that world lines are often impossible to extend to sufficiently unruly and unfamiliar worlds.

6 Quantifiers and well-definedness

The possibility of undefinedness may *prima facie* seem to affect our quantifier rules, for one of the basic ideas of our approach (and, we believe with Quine, every reasonable approach) is that quantifiers range over well-defined individuals. Hence we apparently must reformulate the quantifier rules by adding a condition of well-definedness. For instance, $(C.E)$ would then have to be replaced (by the following):

$(C.E.)$ If $(\exists x)S[x] \in \rho$, then $S[y] \in \rho$ for at least one proper constant 'y'. If 'x' occurs in $S[x]$ extensionally, then also $(y = y) \in \rho$ and $(\exists z)(y = z) \in \rho$ for the same 'y'.

However, the added condition $(y = y) \in \rho$ is redundant for it is a consequence of $(\exists z)(y = z) \in \rho$. For, applied to the latter, the original $(C.E.)$ yields the conclusion that $(y = u) \in \rho$ for some proper constant ρ. But then $(C. =)$ implies that $(y = y) \in \rho$. Hence we do not have to revise our quantifier conditions after all.

This conclusion is of course entirely natural. In order to exist in a world, an individual must be well defined there.

Even though the need of a well-definedness condition does not affect our quantifier conditions, it will turn out to have repercussions for them when they are extended to higher-order quantifiers. I shall not discuss the matter here, however.

A remarkable feature of our treatment of quantifiers is that we did not have to speak of improper constants at all in them. The regularities which hold for improper constants, for instance, the conditions on which existential generalisation is valid for them, are all automatic consequences of the conditions we have imposed on the role of proper constants. We shall illustrate this observation later, after we have formulated the other conditions needed to deal with model sets and model systems.

It is of course in these quantifier rules that the formal expression of the well-definedness of an individual will have to do most of its real work. The simplicity of the rules just formulated in an argument for the naturalness of the procedure adopted here.

7 Atomic sentences, epistemic conditions and well-definedness

There are other places, however, where there is some interplay between well-definedness and truth-conditions. One of our leading ideas will be that $S \in \rho$ expresses that S is true and well defined in $||\rho||$. Here well-definedness of a sentence S means that all the crucial individuals mentioned in S are well defined in $||\rho||$. This idea can be spelled out more fully by means of the following conditions:

$(C.A.)$ If A is an atomic sentence, $A \in \rho$, and 'x' occurs in A, then $(x = x) \in \rho$.

This condition replaces the usual condition $(C.\text{self} =)$ which allows the same conclusion as soon as 'x' occurs in any member of μ and which is now rejected.

Since the membership of S in $\bar{\rho}$ expresses that S is defined and false, we will also have the following condition:

$(C.A =)$ If A is an atomic sentence, $A \in \rho$, and 'x' occurs in A, then $(x = x) \in \rho$.

Furthermore, apart from quantifier conditions and conditions affecting atomic sentences, we need existence clauses in our epistemic rules, as illustrated by the following simplified version of $(C.K)$:

$(C.K)$ If $(x)KS \in \rho \in \Omega$, then $(\exists y)(x = y) \in \rho$ and $S \in \upsilon$ for each epistemic x-alternative in Ω to ρ.

The other epistemic conditions are to be modified in the same way.

Notice that no 'restricted range requirement' is imposed on our quantifiers. For instance, in

$$(\forall x)KS[x] \tag{6}$$

the quantifier '$(\forall x)$' speaks literally of all world lines, as is shown by the condition $(C.U)$, which allows any proper constant to be substituted for 'x'. If we want to restrict these quantifiers '$(\forall x)$' in (6) to actually existing objects and well-defined objects we have to do it by means of explicit realisation:

$$(\forall x)(x = x) \supset KS[x]). \tag{7}$$

8 Different negations

There is an important complication, however, that is caused by the possibility that individuals (world lines) may be undefined in some world $||\rho||$ in the strong sense that the question of their existence or non- existence in $||\rho||$ cannot be raised. The negation of S can now mean three different things.

They will be expressed respectively by $\sim S$, $-S$ and $> S$, and their truth-tables are the following, where *t*=true, *f*=false, *u*=undefined:

S	$> S$	$\sim S$	$-S$
t	f	f	f
u	u	t	f
f	t	t	t

These different negations can be explained as follows:

(a) $> S$ is the mirror image of S, true when S is false, false when S is true, and undefined when S is undefined.

(b) $\sim S$ says that S is not true, i.e. that it is not the case that S is true and defined.

(c) $-S$ says that S is defined and false.

In addition to the different negations we also have a truth-function $+S$ which asserts that S is defined and true. It has the following truth-table:

S	$+S$
t	t
u	f
f	f

We also have a truth-function asserting $= S$ that S is true or undefined (not false and defined):

S	$= S$
t	t
u	t
f	f

As far as the model set and model system technique is concerned, there are obvious connections between the different negations and membership conditions. The following are the basic relations:

$$\begin{array}{lll} > S \in \rho & \text{iff} & S \in \bar{\rho} \\ > S \in \bar{\rho} & \text{iff} & S \in \rho \end{array}$$

$$\begin{array}{lll} +S \in \rho & \text{iff} & S \in \rho \\ +S \in \bar{\rho} & \text{iff} & \sim S \in \rho \end{array}$$

$$\begin{array}{lll} \sim S \in \rho & \text{iff} & +S \in \bar{\rho} \\ \sim S \in \bar{\rho} & \text{iff} & S \in \rho \end{array}$$

$$\begin{array}{lll} -S \in \rho & \text{iff} & S \in \bar{\rho} \\ -S \in \bar{\rho} & \text{iff} & = S \in \rho \end{array}$$

$$\begin{array}{lll} = S \in \rho & \text{iff} & -S \in \bar{\rho} \\ = S \in \bar{\rho} & \text{iff} & S \in \bar{\rho} \end{array}$$

Moreover, we have the following equivalences:

$$\begin{array}{llll}
\sim\sim S & \text{is equivalent} & \text{with} & +S \\
--S & & \text{with} & =S \\
++S & & \text{with} & +S \\
==S & & \text{with} & =S \\
\\
\sim -S & & \text{with} & =S \\
\sim> S & & \text{with} & =S \\
\sim +S & & \text{with} & \sim S \\
\sim= S & & \text{with} & -S \\
\\
-\sim S & & \text{with} & +S \\
-> S & & \text{with} & +S \\
-+S & & \text{with} & \sim S \\
-=S & & \text{with} & -S \\
\\
+\sim S & & \text{with} & \sim S \\
+-S & & \text{with} & -S \\
+>S & & \text{with} & -S \\
+=S & & \text{with} & =S \\
\\
=\sim S & & \text{with} & \sim S \\
=-S & & \text{with} & -S \\
=>S & & \text{with} & \sim S \\
=+S & & \text{with} & +S
\end{array}$$

These equivalences show that all sequences of our truth-functions reduce to one-member sequences. Moreover, the first list of equivalent conditions shows that the following conditions are enough to cover all cases:

$$\begin{array}{lll}
S \in \rho & & \\
S \in \bar{\rho} & & \\
+S \in \bar{\rho} & \text{or} & \sim S \in \rho \\
-S \in \bar{\rho} & \text{or} & \sim S \in \rho
\end{array}$$

We could even use the following:

$$\begin{array}{l}
S \in \rho \\
-S \in \rho \\
\sim S \in \rho \\
= S \in \rho
\end{array}$$

This shows that we could dispense with the conditions that deal with the complement $\bar{\rho}$ of μ altogether, and return to the original model set technique. However, in the following we shall formulate our conditions mostly either in terms of the two negations—and $\sim$ or in terms of $+$ and $=$.

9 Different negations and well-definedness

Given a sentence S and a world $||\rho||$ described by ρ there are three possibilities:

(i) S is defined and true

(ii) S is undefined

(iii) S is defined and false.

These will be reflected in membership terms as follows:

$$
\begin{array}{lll}
\text{(i)} & S \in \rho & \\
& \sim S \in \bar{\rho} & \\
& = S \in \rho & \\
& -S \in \bar{\rho} & \\
& +S \in \rho & \\
\\
\text{(ii)} & S \bar{\in} \rho & S \bar{\in} \bar{\rho} \\
& = S \in \rho & +S \in \rho \\
& +S \bar{\in} \rho & +S \bar{\in} \bar{\rho} \\
\\
\text{(iii)} & S \in \bar{\rho} & \\
& +S \in \bar{\rho} & \\
& = S \in \bar{\rho} & \\
& \sim S \in \rho & \\
& -S \in \rho &
\end{array}
$$

There is some redundancy in these lists. The following suffices to describe the three cases:

(i) $S \in \rho$

(ii) $\sim S \in \rho, \qquad -S \in \rho$

(iii) $S \in \rho$

The truth-tables for negated and unnegated conjunctions and disjunctions are easily set up. Then it can be seen that De Morgan's laws hold for each of the two negations $-$ and $\sim$, i.e. the following pairs of sentences are equivalent:

$$
\begin{array}{lcl}
-(S_1 \& S_2) & \text{and} & -S_1 \vee -S_2 \\
-(S_1 \vee S_2) & \text{and} & -S_1 \& - S_2 \\
\sim (S_1 \& S_2) & \text{and} & \sim S_1 \vee \sim S_2 \\
\sim (S_1 \vee S_2) & \text{and} & \sim S_1 \& \sim S_2
\end{array}
$$

By using these equivalences, the two negations, $-$ and $\sim$, plus any sequences of such negations can always be pushed deeper into our sentences over conjunctions and disjunctions.

Hence we do not need any new conditions or model sets for conjunctions and disjunctions.

The same possibility of pushing negation-signs deeper is easily seen to hold also for quantifiers. In other words, quantifiers don't cause any new work, for it is easily seen that the following are equivalent:

$$\begin{array}{lcl} -(\exists x)S[x] & \text{and} & (\forall x) - S[x] \\ -(\forall x)S[x] & \text{and} & (\exists x) - S[x] \end{array}$$

$$\begin{array}{lcl} \sim (\exists x)S[x] & \text{and} & (\forall x) \sim S[x] \\ \sim (\forall x)S[x] & \text{and} & (\exists x) \sim S[x] \end{array}$$

Hence we don't need any quantifier conditions besides those already formulated, for we can push negation-signs deeper so as to be within the scope of a quantifier.

10 Atomic sentences and well-definiteness

Another place where the possibility of indefiniteness impinges on our conditions and rules is in connection with atomic sentences. We must formulate conditions which regulate the interplay of negation and self- identity in connection with atomic sentences.

First, the same general situation obtains for an atomic sentence A as was found above in Section 9 to obtain for an arbitrary sentence S. The only difference is that now we cannot any longer push negation-signs deeper into our sentences. Hence we have to consider the same three cases (i)–(iii). For an atomic sentence A and require that only one of these three cases obtains. This is most naturally done in stages.

(C.comp) If $S \in \rho$, then not $S \in \bar{\rho}$

We can limit this and the following conditions to atomic sentences with or without the prefix $+, =, \sim$ or $-$.

(C.dual)
$$\begin{array}{lcl} +S \in \rho & \text{iff} & \sim S \in \bar{\rho} \\ = S \in \rho & \text{iff} & - S \in \bar{\rho} \\ +S \in \bar{\rho} & \text{iff} & \sim S \in \rho \\ = S \in \bar{\rho} & \text{iff} & - S \in \bar{\rho} \end{array}$$

(C. neg)
$$\begin{array}{lcl} S \in \rho & \text{iff} & + S \in \bar{\rho} \\ S \in \bar{\rho} & \text{iff} & - S \in \rho \end{array}$$

($C.A$) for any atomic sentence A, one and only one of the following holds:

1. $A \in \rho$

2. $\sim A \in \rho, -A \in \bar{\rho}$
3. $A \in \bar{\rho}$

From $(C.A)$ it follows that at least one of the three sentences $A, -A, \sim A$ must occur in each model set ρ for each atomic sentence A in the language of ρ. No harmful consequences ensue from this, however, for (a) the language may sometimes contain only a finite number of proper constants. Then ρ need not even be infinite. Also, (B), it will be seen that the presence of the sentences $A, -A$ and $\sim A$ does not necessitate the presence of any other sentences in ρ, except for those that must be there because of $(C.=)$.

The interplay between atomic sentences and well-definedness can obviously be regulated by the following condition:

$(C.A =)$ If $S \in \rho$ or $S \in \bar{\rho}$ and if x occurs extensionally in S, then $(x = x) \in \rho$. Here S can be restricted to be of the form $A, +A, = A, -A, \sim A$.

$(C.A =)$ (1) If $= A \in \rho$ and if $(x = x) \in \rho$ or $(a = a) \in \rho$ for each constant x or a in A, then $A \in \rho$.
(2) If $\sim A \in \rho$ and if $(x = x) \in \rho$ or $(a = a) \in \rho$ for each constant x or a in A, then $-A \in \rho$, and $A \in \bar{\rho}$.

Instead of $= A \in \rho$ we can of course have here $-A \in \rho$.

This takes care of the treatment of definiteness. As the reader can see, it affects the model set technique essentially in only one way: by necessitating the modification of the conditions for atomic sentences.

11 Epistemic conditions

It remains to reconsider the conditions needed to govern the concept of 'knowing that'. They do not cause any problems, however. All we have to do is to formulate conditions analogous to the usual ones for all the different sequences of negation-signs that we need. It follows from what was said in Section 8 of the present paper above that we need only four rules, which will be called $(C.K)$, $(\bar{C}.K)$, $(C. \sim K)$, and $(\bar{C}. - K)$. They can be formulated as follows:

$(C.K)$ and $(\bar{C}.K)$ and $(C. \sim K)$ as usual.

$(\bar{C}.-K)$ If $-(x)KS\bar{\in}\rho, \rho \in \Omega$, then $-S \in \bar{\rho}$ for each epistemic x-alternative $v \in \Omega$ to ρ.

The import of these conditions will become clearer if we introduce a number of defined notions.

$$(x)PS =_{\text{def}} \sim (x)K \sim S$$
$$(x)K_wS =_{\text{def}} (x)K = S$$

The meaning of these is obvious. Whereas $(x)KS$ can be true in $\|\rho\|$ only if S is determined and true in each epistemic x- alternative to $\|\rho\|$, $(x)K_wS$ is true as soon as S is true if defined (true or undefined) in each such alternative. Thus K_w

expresses a *weak sense of knowledge* which does not exclude indefiniteness. It says merely that S is *true if defined* in each of the knower's epistemic alternatives.

What $(x)PS$ expresses is that x's knowledge allows S to be defined and true. This clearly amounts to *epistemic possibility*.

The new notions are interdependent. For instance, the following are equivalent: (14.1) $> (x)PS$ (14.2) and $(x)K_w > S$.

In fact, $(C. \sim K)$ and $(\bar{C}. - K)$ could be replaced by the following:

$(C.P)$ If $(x)PS \in \rho \in \Omega$, then $\sim S \in \lambda$ for at least one epistemic x-alternative $\lambda \in \Omega$ to ρ.

$(C.K.)$ If $-(x)K_wS \in \bar{\rho}, \rho \in \Omega$, and if $\mu \in \Omega$ is an epistemic x-alternative to ρ, then $-S \in \mu$.

12 Different Kinds of Knowledge

The power and subtlety of our approach is shown by the fact that it enables us to make several distinctions which are relevant to the role of the concept of knowledge in everyday language and in philosophical discussions. Here we shall consider two kinds of distinctions:

1. Between different kinds of knowlege 'about everything'
2. Between different kinds of knowing who (what, where, when, etc.).

We can distinguish at least the following different kinds of knowledge 'about everything':

(i) Knowing something about each actually existing individual:

$$(\forall x)(\exists y)(y = x \& KS[y]). \tag{8}$$

(ii) Knowing that something is true of all individuals:

$$K(\forall x)S[x]. \tag{9}$$

(iii) Knowing something about all individuals, no matter whether they exist in the actual world or not:

$$(\forall x)KS[x]. \tag{10}$$

The logical relations between these are as follows:

(iii) entails (ii);

(ii) does not entail (iii);

(i) does not entail (ii) or (iii).

(ii) does not entail (i)

(iii) entails (i).

As a simple example concerning the interrelations of (i)–(iii), we shall prove that (iii) logically implies (i):

1. $(\forall x)KS[x] \in \rho \in \omega$ (premise)
2. $(\forall x)(\exists y)(y = x\&KS[y] \in \bar{\rho}$ (counter-assumption)
3. $(\exists y)(y = z\&KS[y]) \in \bar{\rho}$ (from 2. by $(C.U)$)
4. $(\exists y)(z = y) \in \rho$ (ibid.)
5. $z = y \in \rho$ (from 4. by $(C.E)$)
6. $z = z \in \rho$ (from 5. by $(C.A)$)
7. $KS[z] \in \rho$ (from 1. by $(C.U)$)
8. $(z = z\&KS[z]) \in \bar{\rho}$ (from 3. and 4. by $(C.\bar{E})$)

First branch:

9. $(z = z \in \rho$ (From 8. by $(\bar{C}.\&.)$)

Here 6. and 9. violate (C.comp), closing this branch.

10. $KS[z] \in \bar{\rho}$ (from 8. by $(\bar{C}.\&)$)
11. $S[z] \in \bar{\lambda}$ (λ is an epistemic alternative to ρ; from 10. by $(\bar{C}.K)$)
12. $S[z] \in \lambda$ (from 7. by $(C.K)$)

Here 11. and 12. violate (C.comp), closing the second branch and concluding the proof.

13 Example: Self-knowledge

Even though the distinction between K and K_w may seem small, it can surface in ordinary usage and in philosophical puzzles. From the distinction, it follows that there are two senses of b's *knowing who* (or *what*) d is, expressed respectively by

$$(\exists x)(b)K(d = x) \tag{11}$$

and

$$(\exists x)(b)K_w(d = x) \tag{12}$$

The former says that d is defined and one and the same individual in all of b's epistemic alterantives. The latter says that b is the same individual in all those b's epistemic alternatives in which it is defined.

From this further consequences ensue. For instance, while (11) and

$$(\exists x)(x = d) \mathbin{\&} (b)K(d = s)) \tag{13}$$

are equivalent, assuming that d actually exists, (12) and

$$(\exists x)(x = d \mathbin{\&} (b)K_w(d = x)) \tag{14}$$

are not, even on the assumption that $(x)(d = x)$. The reason is clear: (12) says that there is a world line which agrees with b wherever it is defined among b's knowledge-worlds. But (14) adds to this the further information that one of such world lines extends also to the actual world. Hence, we can happily have, for some model system Ω and a model set ρ.

$$\begin{gathered}(\exists x)(b)K_w(d = x) \in \rho \in \Omega \\ (\exists x)(x = d \mathbin{\&} (b)K_w(d = x) \in \rho.\end{gathered} \tag{15}$$

In fact the former requires that, for some proper constant 'z',

$$(b)K_w(d = z) \in \rho \tag{16}$$

But since 'x' does not occur extensionally in '$(b)K(d = x)$', we don't need to have $(z = z) \in \rho$. Indeed, we can consistently have

$$(z = z) \in \rho \tag{17}$$

and

$$(\exists t)(z = t) \in \rho. \tag{18}$$

This observation turns out to be relevant to a minor philosophical puzzle. Hector-Neri Castañeda has presented an ingeneous example (involving an amnesiac war here *incognito*) who comes to know all about the exploits of the hero without realising that he himself is the hero. Calling him 'h', the following are then true in Castañeda's story:

$$(\exists x)(h)K_w(h = x), \tag{19}$$

that is, b knows in the weak sense who b is, and

$$\sim (\exists x)(x = h \& (h)K_w(h = x)), \tag{20}$$

i.e. h does not know that he himself is h. Hence our treatment captures neatly Castañeda's thought-experiment. Notice, in particular, how accurately our semantical conditions match the facts of the story: h knows who h is in the true sense of being able to locate the hero in all his knowledge worlds in which the hero (individual going together with a certain world line) is defined. However, h is not able to trace that world line back to the actual world. Hence Castañeda's example does in fact capture a subtle but very real aspect of the logic of knowledge.

This subtlety nevertheless has nothing to do with the logic of *self-* knowledge, contrary to what Castañeda thinks. For the relation of (19) to

$$(\exists x)(x = h\&(h)K_w(h = x)) \tag{21}$$

is merely a special case of the relation of a more general relation which is not restricted to self-knowledge. This is how it ought to be, however. Even in Castañeda's own story, if a friend of h had come to know (for instance, from reading) the exploits of h so well that she could say,

$$\text{I know who } h \text{ is} \tag{22}$$

the sense in which this would have been true is one in which (22) does not imply knowing of h that he is the war hero. Self-knowledge comes in only through the temptation to say 'but surely the war hero knew who *he* himself was!'. This may be true in some sense endemic to self-knowledge, but that sense is radically different both from the kind of self-knowledge that can be expressed either by K or K_w.

Department of Philosophy, Boston University, USA

9

In Defence of Classical Principles

Imre Ruzsa

1 Semantical versus Syntactical Foundations

In contemporary logic the presentation of a logical theory follows, in general, the line of thought given in (A) and (B) below:

(A) The language (grammar) of the system is described. First the primitive expressions of the language—among them logical constants, auxiliary signs (e.g. brackets), variables bindable by operators, and nonlogical (descriptive) constants—are enumerated, sometimes grouped according to their category (logical type), allowing for extra-categorical signs, too. Then rules are specified by which from certain expression(s) compound expressions can be generated. Generally in these rules it is explicitly stated to which category the input(s) and the output can belong. As a result the class of the well-formed expressions of the language subdivided according to the different categories is defined by an inductive syntactical definition. Among the syntactical categories of the language there is one which is analogous to the category of the indicative sentence in natural language, which is called 'formula', or 'well-formed formula', or simply 'sentence'. A more general method is to define a family of languages instead of a single language, where the members of the family can differ only in the class of non-logical constants.

(B) The relation of consequence (i.e. the relation 'A follows from Γ' where A is a sentence and Γ is a set of sentences) and as a special case the notion of logical truth or validity is defined. The definition of consequence can be either syntactical or semantical. In the case of a syntactical definition we speak about deductibility (A is deductible from Γ), and the relation of deductibility is introduced by an inductive syntactical definition. In a semantical definition of consequence first the class of admissible interpretations is given (most often these interpretations are set-theoretical entities), and it is specified what *semantic value* the well-formed expressions of the language can take. The possible semantic values of sentences are normally called truth values. In the classical case there are two truth values, Truth and Falsity, in other cases it is postulated that among the truth values there is one or more which is a 'distinguished' value. Then the definition of 'A is a semantic consequence of

P. I. Bystrov and V. N. Sadovsky (eds.), Philosophical Logic and Logical Philosophy, 139–150.

Γ' (or '*A* follows semantically from Γ') may be formulated as '*A* is true (*A* bears a distinguished value) in all interpretations where every member of Γ is true (bears a distinguished value)'. This definition may be strengthened by some additional clauses.

When the relation of consequence is defined both syntactically and semantically the issue of the equivalence of the two definitions is raised. In several logical systems the equivalence can be established by soundness and completeness theorems. Most often the definition of deductibility precedes the introduction of the semantics, and then the soundness and completeness of the syntactical system is proved with respect to the semantics. As we shall see presently this order can give rise to some common misinterpretations.

What I have said under (A) and (B) are formal criteria of the logical systems; they are necessary preconditions if we want to have a precise notion of the logical system. But I do not think that they are sufficient preconditions in order to make a formal system a *proper* system of logic. This title can be earned by fulfilling the following non-formal requirements.

(a) The meaning of the logical constants of the language should be translatable into meanings of words in natural language. (E.g. the symbol '&' represents certain uses of 'and' connecting indicative sentences.) This explanation cannot rely on any special field or branch of knowledge, i.e. a symbol is a logical constant if, and only if due to its meaning it can occur in any field as a linguistic tool to structure verbal information.

(b) It should be explained to which type of expressions in natural language the expressions in the different categories (logical types) of the formal language correspond. If the descriptive constants are symbols which have no corresponding meaning in natural language, then we should explain what sort of expressions in natural language they would be suitable to represent. Moreover, if the non-logical (descriptive) constants are meaningful in natural language too, then the relation between the intended meaning in the formal system and the natural language meaning of the symbol should be clarified (often the logical meaning is a restricted, disambiguate derivative of the natural language one).

(c) In general: it should be explained the *logical structure* of which expressions of the natural language can be represented by the different well-formed expressions of the formal expressions of the formal language, perhaps allowing the substitution of certain components of the language by certain—logical or nonlogical—constants of the formalized language.

(d) The definition of consequence—no matter that it is given in syntactical or semantical terms—should correspond to the intuitive notion of the valid inference. This vague postulate can be formulated more precisely by referring to the notion of *truth*. An inference can only be valid if the truth of its premises coerces the truth of its conclusion, and this is effectuated only by the logical structure of the statements and by the meaning of the logical words occurring in them. This necessary condition can perhaps be supplemented by postulates of relevance, e.g. that the conclusion should contain only those non-logical (descriptive) constants which occur in the premises too.

Postulates (a)–(d) together express that a logical system should be accompanied by an intuitive semantics, which substantiates the claim that the formal system is a logical one: it is capable to judge the validity, the correctness of at least some sort of argumentations.

If a formal system satisfies postulates (A) and (B), but is does not conform to stipulations (a)–(d), then it can be call a quasi-logical system. Some of the quasi-logical systems are very remarkable from a mathematical point of view, they can have very important applications, nevertheless they are not theories of consequence, i.e. they are not logical systems *proper*.

If a logical system is supplied with a formal semantics, it obviously has to conform to the informal, intuitive semantics of the system, i.e. the formal semantics can be nothing else as a precise and exact reformulation of the intuitive semantics. Therefore if a system lacks an intuitive semantical explanation then the fact that the system has a formal semantics does not entail that it is a logical system proper. I have hinted at this when I was speaking about the danger of misinterpretation when first a syntactical consequence relation is defined, only after this is a semantics introduced, and finally the soundness and completeness of the calculus with respect to the semantics is proved. With some mathematical invention adequate semantics can be constructed to practically any calculi whatsoever, the existence of a formal semantics in itself does not guarantee the fulfilment of criteria (a)–(d). The semantics of system **R** of relevant logic can be an alarming example, as it contributed nothing to the clarification of the notion of 'relevant implication'.[1]

2 How Many Truth Values are There?

In view of the argumentation in (d) a logical system proper cannot lack the notion of truth. The intuitive criterion of the validity of inference is based on the assumption that true sentences can be distinguished from those ones which are not true—in the classical idiom: the class of sentences can be divided into the subclasses of true and false sentences, which are disjoint and mutually exhaustive.

In classical formal semantics it is established by an inductive definition which formulas are considered true (in a given interpretation). In an intensional semantics this definition usually determines whether a particular sentence is true in a 'possible world': so the truth value of some sentences may change 'from world to world'. If the logical system considers tensed sentences too, the definition of truth will refer to moments of time as well. So generally we can say that in a formal semantics it is established by an inductive definition which formulas are true in a particular moment of time in a possible world of a given interpretation. (Classical extensional logic then is a borderline case, as there is only a single possible world and only one moment of time.)

[1] R. Routley and R. K. Meyer, The semantics of entailment. In: H. Leblanc (ed.), *Truth, Syntax and Modality*, pp. 199–243, Amsterdam, 1973.

This formal semantics is sometimes attacked, that the True/False dichotomy does not hold 'in reality', the notion of a true sentence is vague, among other things because most monadic predicates do not have such a clear-cut extension (truth domain) which would be presupposed by a formal semantics. Consider e.g. predicates like 'is beautiful', 'is clever', 'is a rich man' etc. Some would refer to these cases in order to argue for the inevitability of a 'fuzzy logic'. I would not disclaim the use and applicability of fuzzy logic in say medical diagnostics or in production management, but this is no motivation for calling it a theory of logic. Consider the following two sentences which contain 'vague' predicates:

(1) Agatha is beautiful and clever.

(2) Agatha is clever.

According to classical logic, (2) is a logical consequence of (1): if (1) were true, (2) would have to be true too. Whether we can assign a truth value to (1) 'in reality' is indifferent to a logical theory. In a formal interpretation a truth value can be attributed to (1). And when somebody states or accepts (1), he/she too attributes a truth value to it. If someone accepts (1) while rejecting (2), it is reasonable to accuse him/her of inconsistency. If fuzzy logic holds to the contrary, it is unacceptable. If, on the other hand, it says the same, the allusion to vague predicates is not an argument for accepting it as an alternative or rival logical theory.

Is there some justification to introduce more than two truth values? Later I will argue that at least in intensional semantics it is recommended to allow a truth value gap, i.e. that some sentences in some world lack a truth value. However a truth value gap, just as its name suggests, is not a third truth value (even though in a formal semantics it can give rise to this impression). It is not an intermediate case between truth and falsity and it is not a special case of either of them.

In many valued logical systems one can find no explanations which on the one hand would be clear and acceptable to intuition, and on the other hand would be strictly observed by the formal semantics. The motivation for introducing more truth values than two could be the distinction of the grades of truth and falsity (e.g. logical truth, mere contingent truth). There are other means to cope with this problem, and there is no example for a many valued logical system where such distinctions were observed strictly.

Sometimes a many valued system serves the end of rejecting some inferences which are accepted in a two-valued logic. It may be the case that the rejected inference is rightly criticized. In these cases, however, a many valued logic is at most a palliative treatment, as there is no intuitive evidence—no explanation of the extra truth values. What we would need instead of palliative treatment is a radical and critical revision of semantic principles.

In the intuitive elucidation of consequence relations only the notion of truth is featured (and falsity, its complementary, and perhaps a truth value gap). There is no intuitive motivation for more truth values. For this reason a multivalued system should be classified among quasi-logical formal systems, which does not challenge their mathematical relevance and technical use.

3 Semantic Value Gaps

In recent discussions truth value gap, or more generally, semantic interpretations which allow the lack of some semantic values, occur quite often. The major cases of semantic value gaps are:

(i) An individual name lacks denotation.

(ii) A (monadic) predicate is not defined (neither true nor false) for some objects. In general: the semantic values of the functors are partial functions (on a corresponding domain).

(iii) Some sentences lack a truth value.

Semantic value gaps are under certain circumstances hereditary from expressions to the complex expressions containing them. In formal semantics the value gap is ordinarily filled up with some special objects, called zero entities. If, e.g. the truth and falsify are represented by numerals 1 and 0 respectively, then the truth value gap is represented by the numeral 2.

There are two conceptions to motivate the introduction of value gaps in a logical semantics: I shall call them the *epistemic* and the *ontological* approach, respectively.

According to the *epistemic* approach semantic value gaps are caused by the incompleteness of our knowledge. All individual names denote something, but we sometimes do not know what it is. All (monadic) predicates are defined for each object (within their domain), but in some case we do not know whether a predicate is true or false for a given object. This approach will determine the rules of the hereditariness of value gaps. E.g. a conjunction will have the value 2 only if one of its members has the value 2 and the other member is not false (i.e., its value is either 1 or 2). By the expansion of our knowledge value 1 and 0 cannot change, value 2, however, can change into 0 or 1. (This is called the principle of monotonicity.) The rule determining the truth value of universal quantification is analogous. The value of '$\forall x.F(x)$' is

1, if the value of '$F(x)$' is 1 for every assignment of x.
0, if the value of '$F(x)$' is 0 for some assignment of x,
2 otherwise.

As we can see the aim of this logic is to express the structure of our knowledge without explicitly using epistemic operators. In so much it is similar to intuitionist logic, which also wants to express the laws of mathematical provability without the explicit use of extra operators. The same argument holds good against both of them:

the consequence relation must be based intuitively on the notion of truth, and truth must not be replaced here by knowledge of truth or by provability of the statement.

According to the *ontological* approach, value gaps are caused by the fact themselves, independent of our knowledge. There are two major sources of value gap:

(i) An individual name has no denotatum. The most elementary case is when the name is a definite description. In an intensional semantics it may happen that though a name has a denotatum, it is nevertheless absent from the domain of individuals of some possible worlds. This certainly presupposes that the domain of actual individuals of the different worlds may be different. (E.g. the denotatum of the name 'Aphrodite' is present in the domain of individuals of Greek mythology, but is absent from the domain of the actual world.)

(ii) The semantic values of the functors are partial functions. This, among other things, means that e.g. a monadic predicate is true of some objects, false for some other ones, and is not defined (is neither true nor false) for the other ones. It would be hazardous to say that this is also possible *due to the facts* (as it is the case under (i)), but it is reasonable to refer to *the facts of linguistic usage*. It is intuitively obvious that the special predicates of physical bodies cannot be applied to numbers, the special predicates of human beings living in a society cannot be applied to animals, the special predicates of living beings cannot be applied to inanimate objects, etc.

A possible objection: Are there any real criteria for determining e.g. the domain of applicability of, say a monadic predicate? Of course there is no universal criterion, the boundaries of the domain of applicability are vague in linguistic usage, just as the domain of truth can be vague. This, however, does not question that we should use the stipulation of clear-cut boundaries in a formal semantics.

Another possible objection: Even though it is true that some predicates are inapplicable to certain objects, it is easier for a logical system to add the cases of inapplicability to those of falsity. (Value gaps falling under (i) can also be eliminated by appropriate means, e.g. definite descriptions can be eliminated by the Russell–Quine method[2].) It is beyond dispute that a semantics without value gaps is far more simple than the one admitting value gaps, but—as we shall see presently—the latter has logical advantages which are well worth the price of a more complicated formal semantics.[3]

4 The Hereditariness of the Semantic Value Gaps

The rule of the hereditariness of semantic value gaps is very straightforward in the ontological approach: value gaps are always hereditary from constituents to the

[2] See, e.g. B. Russell, On denoting. In R. C. Marsh (ed.), *Logic and Knowledge*, London, 1956; and W. V. O. Quine, *Methods of Logic*, revised edition, 1963, sections 36–37.

[3] Admitting truth value gaps does not conflict with the principle of *tertium non datur*—in the case of the ontologic approach. For, this principle holds for statements, not for sentences. If an indicative sentence is—due to the facts—without a truth value, it does not express a statement, and whenever it expresses—due to the facts, again—a statement, it must be either true or false.

whole in an extensional context. In order to give a precise explication of this principle we need to define the notion of 'extensional context'.

First: sentences and individual names are extensional expressions. These can have (in a given possible world, at a given moment of time) *factual values*: the factual value of a sentence is its truth value, the factual value of an individual name is its denotatum.

Second: functors can be extensional or intensional. As a first approximation let us consider a functor extensional if the factual value of its input (argument) determines unambiguously the factual value of its output (the functor filled in with an argument). An extensional functor in a given interpretation (in a given possible world) can have a factual value. The factual value of an extensional functor is a function—most often a partial one—which injects the set of allowed values of its arguments into the set of allowed values of its outputs; this is normally called the extension of the functor. In formal interpretations, intensional (i.e. not extensional) functors are not supplied with a factual value, only with intensions, a different sort of semantic values. From the above it follows that the argument of an extensional functor can only be an expression which has factual value—i.e. which is extensional. When the functor is filled in with an argument semantic value gaps are hereditary both from the functor and the argument to the complex expression.

Sentence functors of classical logic are obviously extensional in this sense; so throughout their application the value gap is hereditary from the component(s) to the complex sentence. How is the value gap inherited in quantification? In order to answer this question we should set aside the standard explanation of the quantifiers, that quantifiers are operators binding variables. Instead let us interpret quantifiers as functors generating sentences from monadic predicates.[4] In the explanation of predicates we do not have to resort to first-order logic: any functor which produces a sentence when filled in with an expression of a particular type α is a monadic predicate. If F is such a predicate, its universal and existential quantification should be denoted by '$\forall(F)$' and '$\exists(F)$', respectively. After this the rule of hereditariness of value gaps can be applied literally: '$\forall(F)$' and '$\exists(F)$' lack a factual value if (and only if) F lacks a factual value.[5]

It is advisable (though not inevitable) that we should use only a lambda operator as an operator binding a variable, with the prescription that '$(\lambda x.A)$' lacks a factual value if and only if A lacks a factual value for all assignments of x.

It is obvious that in a semantics with value gaps it is well worth using the *descriptor* as a primitive logical functor, which generates individual names from first-order monadic predicates. Let F be such a predicate, and let the description formed from it be expressed by the notation 'I(F)'. The semantic rule attached to it is the standard one: In a given interpretation, in a possible world w the value of '$I(F)$'

[4] See, e.g. A. Church, A formulation of the logic of sense and denotation. In Henle, Kallen and Langer (eds), *Structure, Method and Meaning: Essays in Honor of H.F. Sheffer*, New York, 1951, pp 3–24.

[5] If F is false for some object, then obviously, '$\forall(F)$' must be false. Concerning the truth condition of '$\forall(F)$', there might be different views. A fair condition: let F be true of all objects for which it is defined (assuming, of course, that it is defined for some objects).

is u_0, if u_0, is the only individual in w, for which (the factual value of) F is true; in all other cases '$I(F)$' has no factual value (denotatum) in w. In a logical theory admitting value gaps, descriptions are ineliminable.

It may be useful to introduce a functor for the identification of factual values, for example with the notation '$(A = B)$' stating that expressions A and B belonging to the same logical type have the same factual value. This, too, is an extensional functor; therefore if either of A and B lacks factual value, the value of '$(A = B)$' is 2.

If we want to express within the object language that a certain extensional expression lacks a factual value (in a possible world of a given interpretation), we have to introduce one-place sentence functors which do not pass on the truth value gap. This can be e.g. the expression 'it is true that', and let its application to sentence A be written '$+A$', and let its truth value be 1 if A is true, and 0 otherwise (i.e. when A is false or lacks a truth value). Then let

(3) $-A =_{\mathrm{df}} \sim +A$

be the expression for 'it is not true that A' (provided $\sim$ is the sign for negation for which the truth value gap is hereditary); moreover, for an expression B belonging to any extensional type the formula

$$-(B = B)$$

expresses that B lacks a factual value (provided the sign '$=$' is used for the identification of factual values.

5 Intensional Functors

A semantics admitting value gaps becomes most effective when we are about to produce an intensional logic, when we distinguish in the object language extensional and intensional functors. Then a possible world semantics is most appropriate. In the interpretations we are going to include a set I of indices (generally, $I = W \times T$, where W is the set of the (labels of the) possible worlds, and T is the linearly ordered set of moments of time). This first of all makes the definition of the intensions of extensional expressions possible: the intension of an extensional expression A is a function which is defined on I, and its values are factual values belonging to the type of A. If α is an extensional type and $D(\alpha)$ stands for the set of possible factual values belonging to this type (in the given interpretation), the intension of A is a member of the 'domain of intensions'

(4) $\mathit{Int}(\alpha) = {}^{I}D(\alpha)$.

(I use the expression '${}^{X}Y$' for denoting the set of functions from X into Y). If there are free variables in A then the intension of A may be dependent on the assignment of these variables.

Second we can define the semantic value—the intension—of intensional functors. Suppose that an intensional functor F produces an expression of type α when filled in with an argument of type β, where β is extensional and α can be any type. Let us denote the type of F with '$\alpha;\beta$'. Then the intension of F is a member of the domain

(5) $Int(\alpha;\beta) = {}^{Int(\beta)}Int(\alpha)$,

so it is a function assigning intensions of type α to intensions of type β.

We can produce intensional operators by the lambda operator, if we distinguish between extensional and intensional variables within the extensional types. If v is an assignment of variables, x and ξ are extensional and intensional variables, respectively, of type α, then

$$v(x) \in D(\alpha), \qquad \text{and} \qquad v(\xi) \in Int(\alpha).$$

So if A is an expression of type α and ξ is an intensional variable of type β then '$(\lambda\xi.A)$' is an intensional functor of type $\alpha;\beta$ (here β is extensional, whereas α is any type).

We can introduce variables in the intensional types, too, and the application of the lambda operator can be extended to them; this will enhance the force of expression of our language. (I cannot touch upon this theme here[6]). Instead of enumerating further details of the construction of semantics for intensional logic admitting value gaps, I am going to dwell on some important advantages of the system.

Two expressions of the same type are said to be *logically synonymy* if their intensions are the same (coincide) in all interpretations, in all assignments of the variables. If a component of an expression is replaced by one of its logical synonyms, the outcome will be logically synonymous with the original expression. The analogous law was duly criticized in semantics lacking a truth value gap (Montague's intensional logic[7] included), as in these semantics the notion of logical synonymous is very weak. In a semantics with a value gap, however, the criterion of logical synonymy is very strong, so here the side effects which were thought paradoxical in logical systems without a value gap simply do not appear, and this is a direct consequence of the introduction of semantic value gaps.

In a semantics without value gaps the two members of the following pairs are logically synonymous:

(6) A and $A\&(B\vee\sim B)$,
(7) $A\vee\sim A$ and $B\vee\sim B$

---

[6] See, e.g., I . Ruzsa, Intensional logic and semantic value gaps, *Logique et Analyse*, **29**, pp. 187–206, 1986. A more detailed report on this subject is I. Ruzsa, *Intensional Logic Revisited*, budabest 1991 (published by the author).

[7] R. Montague, Universal Grammar, 1970, and The proper treatment of quantification in ordinary English, 1973. In: R. H. Thomason (ed), *Formal Philosophy: Selected Papers of R. Montague*, Yale Univ. Press, New Haven London, 1974.

$$(8)\ F(a)(b) \vee \sim F(a)(b) \text{ and } F(b)(a) \vee \sim F(b)(a).$$

In a semantics with value gaps (6) and (7) do not hold for arbitrary A and B as it is possible that one of them lacks a truth value, while the other one has one, and this is sufficient for refuting logical synonymy. About (8): if F is a dyadic predicate it may well be the case that it is defined for the pair (a, b), whereas it is not defined for the pair (b, a), so logical synonymy does not hold. (This is in accordance with common sense—'Peter visits or does not visit the House of Parliament' and 'The House of Parliament visits or does not visit Peter' are not synonymous sentences).

6 Strong Consequence

In a logic with value gaps the following *strong* consequence can be defined:

Sentence A is a strong consequence of the set of sentences Γ iff

(i) there exists an interpretation with an assignment of the variables and an index i such that no member of Γ takes the value 2 at i,

(ii) whenever[8]no member of Γ takes the value 2, A does not take the value 2 either, and

(iii) whenever every member of Γ takes the value 1, A takes the value 1 as well.

This definition excludes inference to an irrelevant conclusion, e.g. that from A we could infer $(A \vee B)$ or $(B \vee \sim B)$, for any B[9].

In a logic admitting semantic value gaps two sorts of logical truth are distinguished. A formula (sentence) is *irrefutable* if it is never[10] false (i.e. its value can only be 1 or 2). A formula is *valid* if it is always[11] true. Most of the classical logical truths are irrefutable, though not valid. A purely extensional formula[12] can only be valid if it does not contain descriptive (nonlogical) constants, free variables included. Moreover irrefutable formulas will not be pairwise synonymous.

7 Conditional Statements

Classical logic was criticised excessively because material implication was thought to be appropriate to express the logical structure of conditional statements and there have been many unsuccessful attempts to avoid the 'paradoxes' involved. I am convinced that some of the uses of 'if . . . then' can be expressed correctly by material conditional (e.g. 'If Peter leaves on Monday, then Mary will arrive on Tuesday.'), but the most common usage can be reconstructed only in an intensional logic admitting semantic value gaps. I propose the following rule of truth for conditional statements:

---

[8]That is: given an interpretation, an assignment of the variables, and an index where. . .

[9]The notion of a week consequence might be the classical one: the truth of the premises must exclude the falsity of the conclusion (it is not excluded the conclusion without a truth value).

[10]'Never', i.e. at no indices of any interpretation with no assignment of the variables.

[11]'Always', i.e. at all indices of all interpretations and all assignments of the variables.

[12]I.e., a formula involving no intensional functors.

'If A then B' is without a truth value iff A is impossible or '$\sim B$' is impossible.

If both A and '$\sim B$' are possible, but '$(A\& -B)$' is impossible then 'If A then B' is true.[13]

In the remaining cases 'if A then B' is false.

In an exact logical reconstruction we will have to elucidate the meaning of the term 'possible'. For this we will have to introduce a reflexive relation on the set of indices (a relation of alternativeness or accessibility). Then 'A is possible' (in symbols: '$\Diamond A$') means at index i that A is true at some index j such that j is accessible from i (j is an is an alternative of i). The relation of accessibility can be different for conditional statements differing in their subject matter. So we use '$A \rightarrow_R B$' for the logical rendering of 'if A then 'B', where the subscript 'R' refers to the accessibility relation. In this approach 'if' is not a pure logical connective—it is dependent on the interpretation, and it can have various meaning according to the subject matter. Nevertheless some logical laws can be stated about this sort of conditional:

> The rule of detachment and the rule of syllogism (transitivity) hold:
> B is a strong consequence of '$A \rightarrow_R B$' and A;
> '$A \rightarrow_R C$' is a strong consequence of '$A \rightarrow_R B$' and '$B \rightarrow_R C$'

provided sentences A, B, C are not always without a truth value.

However the rule of contraposition does not hold: it may occur that '$A \rightarrow_R B$' is true but '$\sim B \rightarrow_R \sim A$' is false.[14] And this is not a defect of the above definition. Consider the following example:

(9) If John's wife is beautiful, his friends envy him.

(10) If John's friends do not envy him, his wife is not beautiful.

Now, someone may accept (9) and reject (10), saying that the antecedent of (10) does not entail that John is married at all, and hence the truth of the antecedent in (10) is compatible with the non-truth of the consequent, even in the case (9) is accepted.

Finally let us note that the iterated applications of '$\rightarrow_R$' are meaningless.

Apology

The author of this short note thinks that present and future development of logic is possible not by rejecting the principles of classical logic, but by a careful reevaluation and reformulation of them, e.g. by accepting semantic value gaps. Due to the small compass of this note I could not expound all the arguments in details, but I

[13] Note the strong condition: the truth of A together with the non-truth (i.e. both the falsity and the truthvaluelessness) of B must be impossible.

[14] However, '$-B \rightarrow_R -A$' follows from '$A \rightarrow_R B$'.

hope what I have said above was enough to show the relevance and importance of the subject.

Eötvös Loránd University, Budapest

10

The Semantics of 'All A's are B's'

Julius M. Moravcsik

Laws and generalizations may not be the last work but are certainly the first word in our efforts to give rational accounts of reality. Our abilities to formulate and understand these are partly constitutive of our rationality. The applicability of generalizations to reality is fundamental to our belief that there is order in the universe.

The simplest and most general form for laws and generalizations is: 'All A's are B's'. With sentences of this form we can express links between properties, classes, or collections of particulars. Logicians and philosophers have always wanted to find a uniform semantics for a sentence-form of such basic nature. According to the logician's ideal one should be able to look at a collection of sentences with only the logical vocabulary, such as quantifiers, truth-functional connectives, etc. playing roles in the interpretation, and on the basis of such schematic understanding specify the deductive potential of any such collection. This is the original and clearly commendable source for the notion of logical form.

The interpretation of the logical vocabulary presupposes that the syntax of the sentences in which they are embedded be available. We need to know, for example, which strings are sentences, which are the non- logical descriptive phrases and expressions, how are the scopes of the quantifiers indicated, etc. We also need to know formation rules that distinguish combinations like 'red apples' from ones like 'reads slowly'. This allows us to interpret the first as a case of intersection of classes, while assigning a different reading to the second.

Logical form needs to be distinguished from ontological form, or analysis. For example, a sentence like: 'John smokes cigarettes' can be interpreted as committing us to the property of smoking, or to the event of smoking, as well as to the particular concrete entity John, or any subset of these. In the logical analysis of 'All A's are B's' we concern ourselves only with the schema: 'All ... are ... ' and leave the ontological analysis of what fits into the A- and B- slots open.

Logical form in the minimal sense is, then, the representation of sentences within a logically transparent syntax and the interpretation of the

P. I. Bystrov and V. N. Sadovsky (eds.), Philosophical Logic and Logical Philosophy, 151–164.

logical vocabulary. This notion is enlarged when meaning postulates or other devices specifying the contents of various descriptive terms are added. One can also add other factors that occasionally contribute to the determination of the deductive potential of sets of sentences. But the more such factors we add, such as indices for evaluations of sentences at times, places, pragmatic contexts, etc. the less intuitively clear it is that we are dealing with 'logical form' in contrast with just content of any kind. 'He is eating an apple, therefore he is eating fruit' is not an inference of logic, but one of content. On the other hand, 'All humans are mortal and Jones is a human, therefore Jones is mortal', is an inference of logic, since it rests only on syntax and on the logical vocabulary.

Matters are further complicated when some logicians define logical form not directly on the language which they study, but on, e.g., an 'intentional logic' into which the language is translated.[1] This raises the question of what the criteria of adequacy for such an enterprise are. If we invent two artificial languages, then the translation between these is a purely conventional manoeuvre. Alternatively, if we translate from one natural language to another, there are empirical criteria of adequacy that yield at least partial ordering of proposed translations. But the task of translating a natural language into an artificial one calls for criteria not furnished by either of the aforementioned enterprises.

Fortunately, we are dealing with a fairly straightforward issue. There are sentences of English and of other Indo-European languages that have the form: 'all A's are B's (with suitable translations among the languages) where 'A' and 'B' stand for predicate phrases which can have denotations, or ranges of applications. These sentences have inferential potential. We shall consider interpretations of the truth-value of such sentences under the condition in which the subject expression fails to denote.

If such sentences have a logical form, then we should find a uniform interpretation of these. The sceptic, however, would try to convince us that such a uniform interpretation is impossible, and that the specific meanings of the predicate phrases to be used will influence even what we called the logical inference potential.

Three important proposals have been made on behalf of the uniform logical interpretation for such sentences. We shall call one the 'down-to-earth' approach. According to this view for sentences for our scheme, the subject expression carries existential import. Thus if there are no A's, the sentence is false. A second interpretation is provided by modern logic. According to this view the subject expression does not carry existential import, and the sentence is false only if there is an A that is not a B. A third interpretation is due to P. F. Strawson. According to him the subject

[1]E.g. R. Montague. Universal Grammar. In R. Thomason, ed., *Formal Philosophy*, Yale University Press, New Haven, 1974.

expression does not carry existential import, but existential presupposition. Thus if there are no A's, the sentence is without truth- value, or more precisely, the question of its truth-value does not arise.[2]

In this paper I shall argue for two main points. First, that we need a combination of two of these proposals to account for the facts adequately. Secondly, that such a 'mixed' solution requires us to rethink the relation between logical form and the semantics of natural languages. A uniform interpretation of logical semantics will not explain the facts. On the other hand, not anything goes. Between the two extreme positions regarding logical form that were mentioned above, there is room for a golden mean.

1 A Story and Its Interpretation

Consider the following story.[3] A door-to-door salesman is selling toys in a neighbourhood. As he makes his 'pitch' with yet another housewife he says to her: 'But all other children in this neighbourhood have this toy'. The women answers: 'You are a liar. There are no other children in this neighbourhood. Our daughter is the only one'.

This exchange seems reasonable. But, as we saw, it clashes with the interpretation of modern symbolic logic according to which if the subject expression fails to refer or denote, the universal generalization in which it is embedded is true. On the other hand, the 'down-to-earth' camp would support the housewife. Within this conception the subject expression does carry existential import, hence the salesman's statement should be regarded as false.

Let us try the story again, with the housewife playing, successively, the roles of having been convinced by the three interpretations. Here is the first scenario. Salesman: 'But all the other children in this neighbourhood have this toy'. Women: 'Well that is true, but since there are no other children in this neighbourhood, all you are saying is that there is no child around here without this toy. That is a very weak reason for me to make a purchase'.

The second scenario. Salesmen: 'But all the other children have this toy'. Woman: 'Sir, the question of the truth-value of your assertion does not arise, since the subject expression fails to denote'. Or, more plausibility and still within the spirit of Strawson's proposal: 'I don't understand what you mean. There are no other children in this neighbourhood'.

Finally, the 'down-to-earth' scenario. This is the same as the initial description of the scene, sketched above. The response of the 'down-to-earth' housewife is: 'You are a liar. There are no other children in the neighbourhood'.

---

[2] P. F. Strawson. 1963. *Introduction to Logical Theory*, Methuen, London, 1963.

[3] I heard this story many years ago in a class conducted by Manley Thompson. I wish record my gratitude for his inspiring teaching.

My intuitions tell me that the third scenario is the right one, and that the first is the most implausible one. I cannot imagine a parent who would say *in this context* that what the salesman said was true. Nor can I imagine that a salesman who knows and accepts the modern logical interpretation and expects his audience to do the same, would be using this as an intended convincing argument. And yet we all know that comments of this sort are used by salesman, and that in many contexts people do not object to them as illegitimate or pitifully weak. If both salesman and housewife are enlightened, or corrupted—depending on one's point of view—by modern symbolic logic, then the conversation will never take place.

The second scenario does not sound much more plausible that the first, even on the more informal reading. For it seems that the parent knows perfectly well what the salesman 'means'. She knows what the facts are. Since she is dealing with a door-to-door salesman, it is reasonable for her to assume that he too knows what the facts are, especially as far as the number of children is concerned who live, or rather do not live, in the houses the man already visited. Thus puzzlement seems less appropriate than indignation expressed by the housewife.

So far, then, the third solution seems to be winning. But, to rephrase as Aristotelian saying: 'One example a (philosophic) summer does not make.'[4] Let us consider a different story. Let us suppose that MENSA, an international association for people with high IQ's, sponsors an academic competition of some sort for high school students in a certain region. circulars are sent out to every high school, and these are read by the principals to the students. There are qualifying rounds for the competition, but not for everyone. For, as one of the sentences of the circular reads: 'all students with an IQ of at least N (where this is a very high number) automatically qualify'.

My intuitions about this example are quite different from the ones concerning the first one. Suppose that a student who has done research on the IQ's of all of the students in the high schools of the region gets up and shouts: 'You are a liar. There are no students with IQ's that high in any of the schools of this region'. This would be an odd response; squelched appropriately by something like 'That is too bad. But what I said is still true. Alas, no student will qualify automatically'. The Strawsonian response in this case seems just as inappropriate as the "down-to-earth" one. The truth-value of what is asserted in the circular does not vanish because the subject expression fails to refer. So it seems that in this case modern logic carries the day. Yet both sentences have the same, 'All A's are B's' form.

In order account for the clash in intuitions, let us consider the condi-

[4] Aristotle. *Nichomachean Ethics* e.g. in R. McKeon, ed. *The Basic Works of Aristotle*, Random House, New York, 1941.

tions under which it is reasonable to assert the two sentences respectively. The conditions seem to differ. The second sentence expresses one of the rules of the contest. As such, it is designed to cover actual and possible cases, and does not require for its truth or reasonableness previous empirical investigations of the intelligence of the students. It is not unlike the well known sign: 'All trespassers will be prosecuted'.[5] This sign does not require actual trespassers for its force. So its seems that the logical account fits rules and their consequences. But perhaps the key point is not whether we consider sentences that originated as rules, but that the sentence has counterfactual force. In this respect it is similar to laws of nature, both to conceptually and to physically necessary ones. There are laws governing certain domains; for example substances with high temperatures. We can derive from these more narrow laws that hold for substances with extremely high temperatures, and such laws will hold even if it turns out that there are no substances with temperatures that high. We can formulate such laws and have confidence in them even if at the time of formulation we do not know whether the subject expression denotes or not.

Lawlikeness is not restricted to scientific discourse. Within a community shared experience can justify a number of common sense laws, such as 'truly altruistic people do not receive their rewards'. This statement would not be regarded false even if the best psychologists in the world came up with a proof that there are no such altruists.

Our first sample sentence, however, cannot bear this interpretation. Given what we know today about toys, neighbourhoods, and human nature, nobody would think that there is a lawlike connection between being a child in that neighbourhood and having that kind of a toy. But if what the salesman says is not lawlike, then it should be based on the examination of at least a sample of the collection of entities denoted by the subject expression, namely children in the neighbourhood. A rational generalization is based either on discovered laws, or on inductive samples. Given that the community knows that the sentence about all the children in the neighbourhood cannot be taken as a law, its meaning will be interpreted as: 'all the children that happen to be living in the neighbourhood do in fact have ...', and as such it will be judged false.

We found two different interpretations for two different kinds of sentences. The relevant intuitive differences are related to our general conceptions of laws and inductive generalizations. Our next task is to see if this contrast can be articulated better, and if it can be used to account for all of the relevant data.

[5]As J. O. Urmson pointed out in conversation, the 'all' here is at most optimal. Without it we get a bare plural subject, and this is such contexts indicates either a law or regularity.

2 The Lawlike and the Non-lawlike

In this book *Fact, Fiction, Forecast* Nelson Goodman provides a conceptual explication of the distinction between lawlike and non-lawlike sentences.[6] Goodman's main purpose is the adequate analysis of the problem of induction, and of counterfactuals, but in the course of carrying out this analysis he sheds light also on the lawlike vs. non-lawlike distinction. He does this by developing the technical notion of projectability. We shall assume the soundness of his analysis, and use his version of the lawlike—non-lawlike dichotomy. Goodman's example of a universally quantified non-lawlike sentence is:

1. All of the coins in my pocket on V-day are made of copper.

It is easy to see why this is not lawlike. Given our communally shared knowledge of the world, there is no law connecting being a coin in Goodman's pocket on V-day with the property of being made of copper. We could not add to (1): "Furthermore, if any more coins were to appear in Goodman's pocket on V-day, these too would have to be made of copper." For there is no evidence available to us now on the basis of which such a projection could be made. To be sure, the non-lawlikeness of (1) depends on currently shared knowledge. There is nothing in the mere form of the sentence that would rule it out from becoming—under very strange circumstances - some day a lawlike statement. Being lawlike or non-lawlike are not solely formal properties of sentences. These properties depend, rather, on the relation between the content of the sentences and our current shared knowledge of the world; a relation that can change in time.

I will make use of this distinction within the framework of a rich ontology, including realist attitudes both towards the modalities and towards universals. Goodman, Quine, and others with much more austere ontologies than mine, would object to this. But the views proposed in this paper are independent of my realist preferences in ontology. The view proposed could be expressed also within less Platonistic schemes of philosophy.

In drawing the distinction between lawlike and non-lawlike Goodman's concern is with generalizations that hold over collections of actual particulars and those extend to cover also possibilities, since they are supported by counterfactuals. I shall extend the dichotomy to cover also those that must be true in various senses of 'must'. These include the normative (as in rules about trespassing), the mathematical, the conceptual or analytic, as well as the natural or physical 'must'. The lawlike as used in my analysis need not hold for any actual entities as long as it has as its natural domain a set of possibilities. Thus for lawlike sentences under this wide interpretation it does not matter whether the subject expression does in fact have denotation. Of course, it would matter if it were true of one of these sentences

[6] N. Goodman. 1954. *Fact, Fiction, Forecast*, The Athlone Press, London, 1954.

that the subject expression could not have denotation in the logical sense of 'could'. But then we would deal with self-contradictory predicates, and these are not included under this analysis.

In summary, my notion of a lawlike sentence includes the following (a) Definitional generalizations and their logical consequences that are also in universally quantified form. For example: 'all brothers are siblings'. (b) Explicit rules like the one about trespassers. (c) All universally quantified statements of mathematics. I take it that this discipline deals with necessary propositions ranging over an abstract domain. Thus its sentences cannot be preferred by: 'it happens to be the case that ...'. This characterization does not rule out the possibility of empirical truths about numbers such as 'Jones can think of any of the positive integers between 1 and 100'. (d) The laws of the empirical sciences. (e) Common sense laws about empirical reality.

In all of these cases we are dealing with a part of the epistemic status of a sentence. We are giving this a communal interpretation. Assigning to a sentence the status of lawlikeness or withholding such a rank is not a matter of individual choice. Nor will it do to say that someone could choose to interpret a given sentence in either of these two ways. These interpretations are matters of common sense or science.

Thus they are communal enterprises, and there are undoubtedly many borderline cases. For these cases my proposal predicts that our intuitions about existential import will waver.

Let us consider a few examples. The sentence

2. All furniture located in Africa will be subject to intense heat.

is lawlike, since it is not just about the furniture that happens to be in Africa, but about any possible furniture piece with this location. We can know that (2) is true, without having examined any of the furniture in Africa; in fact we can assert (2) with confidence without knowing whether there is any furniture in Africa. The law is based on our knowledge of the constituency of furniture and on temperature conditions in that part of the world. Let us consider now

3. Financial analysts today know how to analyse data on computers.

In this case our intuitions about existential import waver. This is due to the fact that our conception whether (3) is lawlike wavers. Computers are a new type of artefact, financial analysts make up a new profession; thus one does not know whether to take (3) simply as a non-lawlike statement based on a sample of analysts, or a regulative statement, based on emerging requirements for the profession. Complications arise also when we consider subject expressions that contain reference to specific location. Thus

4. All water is H_0^2.

will be taken by everyone as lawlike, and true even if there was no more water in the world. This analysis will still hold for a sentence like

5. All water in lakes is H_0^2.

But what about a sentence like

6. All water is Lake Michigan is H_0^2?

Some of our intuitions say that this should be derivable from the lawlike (4), and thus lack existential import. But some of our intuitions might suggest that (6) is equivalent to:

7. All of the water that happens to be in Lake Michigan is H_0^2.

The trouble lies in the subject expression. It can be interpreted as a qualitative descriptive phrase, with the locator as a kind of adverbial, or it can be interpreted as a designator, singling out a concrete object in a concrete place. Thus the best solution is to treat (6) as ambiguous; on one reading it is lawlike, on the other it is not.

The main proposal of this paper, emerging from these considerations says that there is no uniform logical semantic interpretation for universally quantified generalizations. There are two kind of such generalizations. The lawlike ones do not have existential import, but the non-lawlike ones do. The interpretation for the lawlike ones is: 'if anything is A, then it is also B'.[7] This holds for all lawlike sentences, empirical or otherwise.[8]

The non-lawlike generalizations say something that happens to be true of a certain particular collection of particular entities, and hence they are about the things that in fact make up the denotation of the subject expression.[9] Thus they do have existential import, and this coincides with what we called above the 'down-to-earth' analysis. The reasonable conditions of assertion for non-lawlike generalizations entail the inspection, direct or indirect, of at least some of the members of the collection denoted by the subject expression, and thus imply that someone has had contact with these, existing, entities.

[7]As E. Morscher pointed out in conversation, problems arise when we consider the same lawlike generalization with the corresponding negative predicate. The problems with positive and negative predicates have deep roots, culminating in the so called 'raven's paradox', they affect equally all approaches to this matter.

[8]This intuition that in the first formulated by me at Stanford-UCLA philosophy of language conference in the early 1970's. It was brought up also—but with a narrower constraint—by Peter Strawson in conversation in the early 1980's. It is recorded in F. de Jong and N. J. Verkuyl, 1987. Generalized Quantifiers: the Properness of their Strength. In J. van Bentham and A. van der Meulen, eds, *Generalized Quantifiers: Theory and Application, GRASS 4*, Foris, Dordrecht, 1987, but as far as I can tell, this paper is the first attempt to construct a full philosophical theory of existential import in U.G., and discuss related matters affecting logical form.

[9]I am indebted to David Widerker for pointing out that on some interpretations what the subject expression refers to is one complex entity.

Some might object by saying that this proposal brings epistemological considerations to bear on matters of semantic interpretation, and that such moves are illegitimate. After all, we classify statements as known a priori or empirically, but we do not make this classification a part of the meanings of the relevant sentences. For example, 'the cat purrs' does not say: 'it is empirical that the cat purrs'. Furthermore, people can disagree about the epistemic status of a truth expressed by a sentence without disagreeing about the meaning of the sentence. For example, according to J. S. Mill, '$2 + 2 = 4$' is empirically true, while according to Carnap it is a priori. But it seems implausible to suppose that Mill and Carnap disagree about what the sentence means.

In reply we can point to important differences between epistemic status such as being a priori or empirical and the kind of theory-bound conceptual status that is represented by the lawlike vs. non-lawlike distinction. Mill and Carnap would not disagree about what can be inferred within simple arithmetic from the sentence considered above. But difference in inference potential, also within a particular science, or part of common sense, is precisely what is a stake in our analysis of generalizations.

The a priori—empirical distinction is not analogous to the lawlike–non-lawlike distinction. A better analogy is provided by the distinction between such modal notions as necessity and contingency. For disagreement about whether a statement is necessary or contingent does affect inference potential. Furthermore, we do regard modal status as a part of the meaning of a sentence; hence the use of a modal operators in logical analyses of sentences. My proposal suggests that the lawlike– non-lawlike distinction is similar to what would be represented by operators within the logical analysis.

This proposal explains the differences we felt in connection with the initial examples of the saleman's talk and the trespassing sign. The salesman's statement is non-lawlike, hence the housewife's response is correct. There is a failure of existential import in a statement that purports to have one. Hence the statement is false. The statement about trespassers, or the rule about the students qualifying for a competition are lawlike—hence they do not carry existential import, and are true even if there is no instantiation.

The proposal might run into difficulties in connection with sentences like

8. All humans are mortals.

This is lawlike, and thus does not carry existential import. But our semantic intuitions might be obscured by the fact that we all assume the existence of humans. Indeed, without that assumption how could we assign this paper to an author, and how could the author hope to reach an audience? But let us project ourselves into the position of a scientist from another planet who worked out biological classifications for different kinds of living things, inhabiting this planet. From that point of view the lawlike interpretation is more transparent; the scientist might have come to the conclusion that there must be a law like (8), even before having had any acquaintance with any humans. It is only our anthropocentric position that obscures for us the lawlike status of (8). On the other hand,

9. All children in this neighbourhood have this toy

does carry existential import, since—as we saw—in a rational community having the knowledge we have of the world today, the statement must be based on inductive sampling.

These examples show also that the distinction between lawlike and non-lawlike is not arbitrary, but systematic, presenting a criterion that can be applied across historical contexts and languages. Hence it is a proper basis for the determination of such semantic facts as existential import.

3 Alternative interpretations

Our approach seems to exclude the possibility that one could base the distinction between having and not having existential import on syntactic criteria. There is, however, a proposal in the recent literature that would offer just such a mark.[10] Johnson-Laird thinks that

10. All of the boxes are clerks.

carries existential import, and he attributes this to the 'all of the ' phrase. One can agree with Johnson-Laird's intuition about (10) without agreeing with his diagnosis. There are many ways of explaining one's intuition about a sentence like (10). Our proposal explains the intuition by pointing out that the proposition expressed by (10) is not lawlike. One could also try to argue that all sentences expressing generalizations in which the subject expression starts with 'all of the' always introduce non-lawlike assertions. Another explanation would claim that all sentences with this syntactic structure in (10) is accidental.

Phrases like 'all of the clerks' are semantically incomplete. One wants to know: which clerks? Further specifications is expected, either in terms of an explicit relative clause, or by contextual information. Some of these completion may be themselves in need of further specification as in

11. All of the students here are hardworking.

If this is uttered at Stanford University, we can arrive at the interpretation according to which it is claimed that every student at this institution is hardworking. Once understood that way, we can see why it is not lawlike. Given what we know about the world, it is always possible for a very intelligent student to enrol at Stanford, and then—while maintaining respectable

[10] P. N. Johnson-Laird and P. Bara. "Logical Expertise as a Cause of Error; a Reply to Boolos, *Cognition*, **17**, 183–84, 1984.

grades - dedicate himself to the 'dolce far niente' life-style. Furthermore, the phrase 'all of the ' is not a crucial factor. The non- lawlikeness is preserved even if we use other expressions in the subject slot such as 'all', 'every', etc. Let us now consider another sentence that has Johnson-Laird's required syntactic form.

12. All of the students over 6'6" can make a team.

This sentence does have the 'all of the ' phrase in proper position but is lawlike. One can imagine a small high school with a basketball team in dire straights. From this and a few other obvious facts one can grasp the lawlikeness and the truth of (12) without knowing whether there are any students in the school who are 6'6" or taller. This shows that the proposed syntactic criterion is neither necessary nor sufficient. We can explain Johnson-Laird's example and other with the same syntactic structure on the basis of the proposal made in this paper.

We shall now turn to semantic or pragmatic proposals. One of these would relate existential import to the 'data base' or particular epistemic states of speaker and hearer. One could then formulate rules of interpretation for different situations. For example, one might suggest that in cases in which the speaker knows the subject expression to be empty, he should not assert the sentence in question, and when the hearer has the same information, she should judge the sentence to be false. This principle seems like an appropriate pragmatic heuristics for non-lawlike cases, but it is not an adequate principle of either semantics or communicational success for the lawlike cases. If certain generalizations follow from theoretical considerations, then it is reasonable to assert them, either in the case in which we do not know if the subject expression denotes, or even when we do know at the time of the utterance that the subject expression fails to denote. We can see this from examples like

13. All creatures on Mars breathe differently from the way all creatures on earth do.

We know (13) to be true on the basis of general considerations of differences in the atmosphere, and other factors affecting the possibilities for various forms of life. Whether there are any creatures on Mars is irrelevant, especially since it is not logically impossible that at some time there will be such entities.

Let us consider another possible maxim. It says that whenever the speaker knows the subject expression to be in fact denoting, she should indicate this. This does not seem correct either. In the non-lawlike cases it is superfluous. For in these cases the sentences assert the existence of entities within the denotion range of the subject expression. If one says: 'all of the students who are taking Plato this term know Greek', it seems silly to add 'and furthermore, there are such students'. On the other hand, if the context is theoretical or simply calls for the invocation of an empirical law, then adding this information may not be always helpful. It may help the researcher who will do empirical testing, but it may be irrelevant to a theoretician. We can see this in such, unfortunate, cases as

14. All sick people in condition C are beyond cure.

This may be important to know even if there are no people right now in condition C.

It is worth stressing that we often do not know when we formulate a law whether the subject expression denotes or not. Still, research goes on, and sentences are evaluated as true or false, and facts emerge in virtue of which we can see a lawlike sentence to be true regardless or whether the subjects purported to be denoted exist or not.

Another alternative could be based on Grice's work on conversational implicitures. According to this proposal we should retain for all universal generalizations the interpretation assigned by modern logic, and account for divergent intuitions by adding to these interpretations certain implicatures. In the case of conjections a Gricean approach would assign as interpretation the truth-functional one, i.e. the conjunction is true if and only if both conjuncts are true. It would then assign to various contexts maxims such as 'when temporal order between the conjuncts matters, indicate this by ordering the conjuncts in the sentence', thus accounting for the felt difference between 'they got married and had a baby" and "they had a baby and got married'. For universal generalizations one could add maxims like 'be specific' and 'be maximally informative'.

Grice's approach may be useful for some semantics puzzles, but it would not solve the problem of existential import for generalizations. This can be shown by the following considerations.

The maxims would suggest that we should try to provide in all cases as complete information as possible. But we saw in several cases that correct interpretation was not a matter of providing all of the relevant information. In the non-lawlike cases existential import is a matter of entailment, not a matter of what information we provide in context.

In the case of many lawlike sentences the possible non-emptiness of the subject range is simply irrelevant. In those contexts being more informative by adding the information that the subject denotation is not empty does not alter the assignment of truth or falsity.

It is important to keep the Gricean maxims informative and non-vacuous. Maxims like 'be specific in appropriate ways' or 'be maximally informative in relevant ways' are so vacuous as to be of no help in explaining the semantic phenomena that we reviewed.

Permeating the discussion of all of these alternatives is the assumption that the particular intentions of the speaker are irrelevant. Suppose we go back to the initial conversation between salesman and housewife and assign to the salesman the following final rejoinder: 'But madam, you should not call me a liar. For I intended what I said about the children as a lawlike statement, with existential import interpreted as in modern logic'. Surely this would be beside the point. As we said before, what can or cannot be taken and interpreted as lawlike is a communal affair, not an individual decision. Communally shared knowledge, be it among experts or within common sense, determines what propositions various sentences express at various stages of history, and not what is in the mind of an individual speaker.

4 Logical Form Revisited

There may be certain disciplines within which constructing universal generalizations as lawlike or non-lawlike is simply irrelevant. If we think of mathematics as a discipline whose objects are a separate abstract domain the members of which exists necessarily, then the contrast we relied on does not surface. Every generalization will be lawlike, but existential import is guaranteed by the fact that all numbers exist by necessity. Issues of incompleteness and undecidability do not affect these matters. On this interpretation mathematics need not concern itself with denotation failures.[11] The problem of denotation failure and existential import, and the relevant distinction between the lawlike and the non-lawlike affect only that part of discourse within which there is a difference between the contingent and the necessary, or—in Quine's terms— between the peripheral and the deeply entrenched.[12]

It seems, then, that existential import in the case of universal generalizations is not dependent on specific descriptive content, but at the same time it is not a matter of logical form in the traditional sense either. Someone might consider this, and simply position formal representations ambiguity for all universal generalizations, and then claim that context will disambiguate between lawlike and non-lawlike reading. But this seems ad hoc. The usual empirical reasons for positing ambiguity do not apply. Many of the sample sentences we considered are not ambiguous according to the intuitions of competent speakers. Furthermore, the lawlike vs. non-lawlike distinction is relevant because the competent speaker-hearer reasons in terms of this and

[11] I am indebted to Jonathan Stavi for helpful discussions; he should not be held responsible for my conclusions.

[12] Interestingly enough, Aristotle's view of what he would regard as genuine sciences corresponds roughly to what was said here about mathematics. He is interested in the logical analysis of what I call in this paper lawlike sentences, but he also has existential import for generalizations; not on the ground of an analysis of lawlikeness, but on ground of his ontological assumption that every genuine science deals with a natural generic collection of entities, carved out from the total realm of existing things.

assigns existential import by considering this dichotomy.

Our result may have further implications for the relation between syntax and semantics of natural languages. In the past two decades many changes were made in generative syntax motivated by the wish to account for such semantic facts as quantifier scope and anaphoric reference. If something as basically a matter of logic as existential import can be shown to be independent of syntactic structure, then perhaps other logical and semantic aspects of language structure can also be assigned to purely semantic interpretations, and grammar can be freed from this burden.[13]

Stanford University, USA

[13]I am indebted for comments to Asa Kasher, Tanya Reinhart and Scott Soames.

PART IV

Non-Classical Logics

11

Conditional Intentions

Ilkka Niiniluoto

1 Introduction: from Imperatives to Intentions

In the Third Finnish–Soviet Logic Symposium, organised in Helsinki in May 1983, I presented a paper with the title *Hypothetical Imperatives and Conditional Obligations* [14]. My main idea was to analyse hypothetical imperatives ('Do p if q!') in analogy with conditional bets, and thereby to argue that the logical form of conditional obligations, created through successful imperatives, can in a natural way be represented by an intensional conditional sentence ('If it were q, then it ought to be p').[1]

In preparing my lecture for the Intention Colloquium, organised by the Philosophical Society of Finland in Turku in January 1989, I started from the idea that *conditional intentions* might be treated in the same way as hypothetical imperatives. In particular, the analogy with conditional bets might be relevant to the case of intentions as well, since the concepts of command and intention are most obviously close relatives to each other. At least we may observe that in most philosophical theories of action imperatives provide an important example of the different ways in which an *intention to do something*[2] may arise:

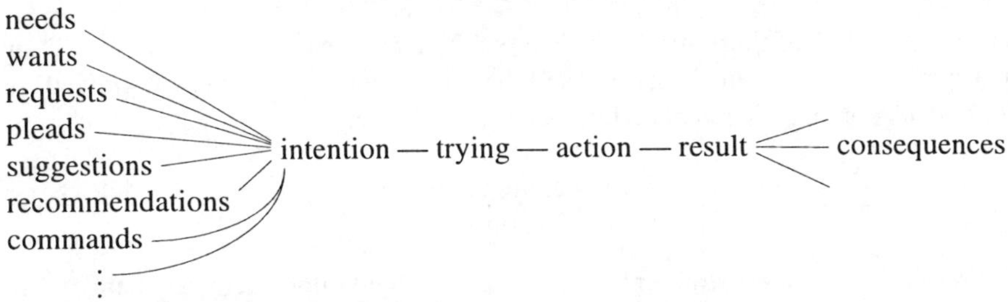

[1] See also my paper [13] for the IVR Congress in Helsinki. For comments on [13, 14], see [8].

[2] In this paper, I am primarily concerned with what Miss Anscombe [1] calls 'expression of intention for the future' and 'intention in action'. But, as intention statements will turn out to be intensional, I am also indirectly concerned with the intentionality of action. For the relation between intentionality and intensionality, see [7].

P. I. Bystrov and V. N. Sadovsky (eds.), Philosophical Logic and Logical Philosophy, 167–179.

(cf. for example, [20]). Further, the debates on the logical form of conditional intentions (cf. [17, 3]) are clearly parallel to the discussions about conditional obligations.[3]

When I started to work out this idea, I first found out (to my pleasure) that Anthony Kenny indeed has explicitly treated commands and intentions as special cases of a more general concept, *fiat* (see [10, 9]). Kenny not only suggests that a 'logic of fiat' is needed; to my surprise, he discusses the idea of analysing conditional commands in analogy with conditional bets—and attributes this idea already to Michael Dummett's classical article 'Truth' [5].

As my basic idea about conditional commands received support from an unexpected direction—viz. from the past—a similar treatment of conditional intentions started to appear ever more promising. In this paper, I intend to report the results of my consequent investigations on conditional intentions. As we shall see, they are in some sense similar, but perhaps in more important senses dissimilar to conditional commands.

2 Intention Sentences

Intentions may be expressed in many alternative ways. For example, the sentences

> I intend to stay at home tonight.
> I am going to stay at home tonight.
> I shall stay at home tonight!

express the same intention. In this paper, I concentrate on sentences which contain the verb 'to intend' in some form. Examples of such *intention sentences* include

> I intend to keep my promises.
> We intend to build a house.
> Vladimir intends to eat fish for dinner.
> Lisa intends to see to it that the children go to school in time.

The subject term of these sentences is typically a personal pronoun ('I', 'we')[4] or a proper name ('Vladimir', 'Lisa'), and the present tense of the verb 'to intend' is followed by an expression describing action or making:

to stay at home tonight
to build a house (1)

Expressions (1) are indefinite relative to persons, and either definite or indefinite relative to time. They are thus descriptions of *act-types* or *generic acts* which may be realised in different circumstances by different agents.

[3] Cf. the literature referred to in [14].
[4] For analysis of 'we-intentions', see [21, 22].

The basic form of an intention sentence might then be represented by

$$a \text{ intends } \varphi \tag{2}$$

where 'a' is a singular name and φ is a description of an act-type. Alternatively, (2) could be written in the form

$$a \text{ intends to do } \varphi^*, \tag{3}$$

where φ^* is obtained from φ by replacing the infinitive (1) by the corresponding gerund (e.g. '(the act of) staying at home', 'building a house').

Castañeda [2] defines *practitions* as expressions obtained from generic act-descriptions (1) by adding to them a singular name:

I to stay at home tonight.
Vladimir to eat fish for dinner. (4)

The logical form of intention sentences is then the intention-operator plus a practition:

Intend [I to stay at home].
Intend [Vladimir to eat fish for dinner]. (5)

In this sense, intention is a *practitional attitude*: practitions have realisation-conditions but not truth-values, and thereby they differ from propositions which have truth-conditions.[5]

It is customary, however, to express intention sentences also in a form, where action is understood in terms of its result, i.e. acting is bringing about some state of affairs. Let p be a sentence which describes a state of affairs (e.g. 'the house is painted red'), so that p may be true in certain circumstances and at certain points of time. Then intention sentences may be written in the following form:

a intends to bring about p.
a intends to make it true that p. (6)

(Cf. [11, 23].)

It is important to observe that intending—unlike wanting and wishing— always involves an action plan of the intending subject. A sentence like

Irina intends to make it true that the house is painted red (7)

is ambiguous, since Irina may intend to paint the house herself ('Irina intends that she paints the house') or she may let someone else do the work ('Irina intends to have the house painted'). As no one may intend for another person, sentences of the form (7) will be understood here to correspond to the schema:

Irina intends: Irina paints the house red.

[5] Important discussions on Castañeda are contained in [18, 19].

The operator

$$a \text{ intends that} \ldots \qquad a \text{ has the intention that} \ldots \tag{8}$$

is denoted by A_a or simply A. Intention sentences are then expressed by

$$A_a p \text{ ('}a\text{ intends that } p\text{')} \tag{9}$$

where p is a sentence describing the action of the agent a. If p is temporally indefinite ('Irina paints the house'), p_t is the corresponding temporally definite sentence ('Irina paints the house at time t').

Castañeda [2] argues that intention sentences cannot be generally reduced to the form (9), since the agent a may be unable to identify himself or herself under the name 'a'. I shall simply assume in this paper that the intending agents don't have this problem of self-knowledge or self-reference, so that intention A_a will be treated as a *propositional attitude* whose scope includes propositions rather than practitions.

Sentences of the form (9) correspond to the *descriptive reading* of intention statements: 'I have the intention that ... ' and 'Peter has the intention that ... ' are true or false sentences, whose truth-values depend on what I and Peter are really intending to do. Thus, they resemble the descriptive sentences about the existence of *obligations* or *duties*:

$$Op \text{ ('it is obligatory that } p\text{', 'it ought to be the case that } p\text{')} \tag{10}$$

where p again expresses the action of some person a, is a true or false description of the obligations that person a has in some situation.

Obligations are normally created through the successful *commands* of a norm-authority (cf. [23]). According to the *prescriptive reading*,

$$!p \text{ ('it is imperative that } p!\text{')} \tag{11}$$

is an imperative which does not have a truth-value. Expressing a command,

Irina, paint the house!

is (in Austin's sense) a speech act, which is not true or false, but may under certain 'felicity conditions' create a duty to Irina.[6] The meaning of such imperative (11) may be analysed in term of their 'execution-conditions' [1] or obedience-conditions.

The descriptive reading of intention sentences (9) should be distinguished from their *declarative reading*. A first-person *expression of intention*, such as

I intend to stay at home!
I intend to become a doctor!

[6] The distinction between descriptive and prescriptive reading of normative sentences is by now standard in deontic logic. See the references in [13, 14].

which may be exclaimed out loud or mumbled quietly in my mind, is a kind of *command directed to myself.* Such an expression of intention is no more true or false than an imperative (11), but again it will have its obedience- and realisation-conditions.[7] In the sequel, such a first-person expression of intention will be denoted by

$$Ip \text{ ('I intend: } p\text{')}, \tag{12}$$

where p is a first-person sentence.[8]

3 Formation of Intentions

The analogy between commands and intentions suggests that the formation of (conscious) intentions can be analysed in the same way as the generation of obligations through imperatives ([14]).

Let $\langle x, y, t_0, !, p\rangle$ be the event that person x commands at time t_0 person y to do p. Suppose p implies that y φ's at time $t \geq t_0$ in place w. If x is a norm-authority for y during the time interval $[t_0, t]$, x does not cancel his or her command for y, and y remains capable of doing φ, then at each point of time $t' \in [t_0, t]$ y has the obligation to φ at t in w.

Similarly, an intention is formed at some time t_0, when a person y expresses (out loud or in his mind) a command Ip to himself. This intention will then remain in force until the time of its realisation, unless it is cancelled for some reason. For example, the truth-conditions for a *temporally definite intention sentence* may be defined as follows: the sentence

$$A_y(y \ \varphi\text{'s at } t)$$

is *true at time* t' if and only if

1. $t_0 \leq t' \leq t$
2. $\langle y, y, t_0, !, p\rangle$
3. p implies that y φ's at t
4. y believes that he is able to φ at t

---

[7] For the analogy between commands and intentions, see [2, 10, 9, 18]. Castañeda represents imperatives in the form

Command[Jones to run],

where the imperative operator is combined with a practition via second or third person; intendings correspond to first person practitions. The command model of action cannot be literally applied e.g. to animals, since the 'brutes' may have intentions that they cannot formulate in language [1]. My paper is thus primarily focused on conscious intentions.

[8] Sellars [17] presents declarative intention sentences with the shall-operator and a future tense statement:

Shall [I will stay at home].

5. y does not cancel or forget his intention before t'.

(See Figure 1.)

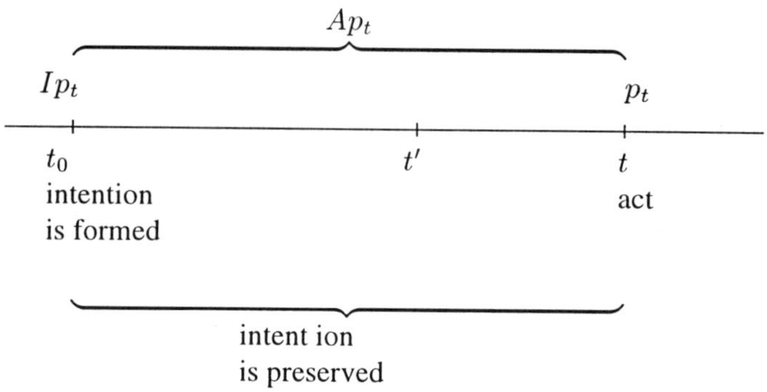

Figure 1:

Similar treatment applies to *temporally indefinite* intentions, which are realised only once ('I intend to travel some day to Novosibirsk'), and *permanent intentions*, which will be realised whenever the occasion is favourable ('I intend to practice every morning').

4 Conditional Imperatives

The realisation of a command or intention is always associated with a *ceteris paribus* condition: the prescribed action will be performed, *if* the circumstances do not change in a decisive way. On the other hand, a command or intention may be attached, already at the time of its formation, with a special condition:

If it rains tomorrow, I shall wear a raincoat!
I intend to meet Vladimir, if I go to Moscow.

Let us denote in the following *hypothetical imperatives* by

$$!(p/q) \text{ ('it is imperative that } p \text{ given } q\text{!')} \tag{13}$$

and the corresponding *conditional obligations* they may create by

$$O(p/q) \text{ ('it is obligatory that } p \text{ given } q\text{')} \tag{14}$$

(cf. (11) and (10).) Similarly, a *conditional intention* in the declarative reading is

$$I(p/q) \text{ ('I intend } p \text{ if } q\text{!')} \tag{15}$$

and in the descriptive reading

$$A_a(p/q) \text{ ('}a\text{ intends that } p \text{ given } q\text{')} \tag{16}$$

(cf. (12) and (9)).

In the debate about the logical form of hypothetical imperatives, Alf Ross [16] and G. H. von Wright [23] have defended the proposal that the conditional command (13) may be expressed by

$$!(q \supset p), \tag{17}$$

where $\supset$ is the ordinary material implication. Then the corresponding conditional obligations would have the form $O(q \supset p)$.

To evaluate this proposal, note that the following imperatives have the same obedience conditions:

$$!(q \supset p)$$
$$!(\sim q \vee p)$$
$$![q\&p) \vee (\sim q\&p) \vee (\sim q\& \sim p)].$$

For example, the command 'If you go to London, don't forget your umbrella!' may be obeyed in two alternative ways: either by staying at home or by travelling to London with an umbrella. Similarly the commands

$$!(q \supset\sim p)$$
$$!(p \supset\sim q)$$
$$! \sim (p\&q)$$

have the same obedience-conditions, as illustrated by the slogans of a Finnish campaign:

If you drive, you will not drink!
If you drink, you will not drive!

which mean the same as

Don't drive and drink (at the same time)!

The proposal (17) for the conditional imperative $!(p/q)$ seems thus natural in the case, where the commanding person wishes to present a genuinely *disjunctive imperative* $!(\sim q \vee p)$, which allows the commanded person to choose between the alternatives $\sim q$ and p [14]. In this case, the condition q belongs to the scope of the imperative operator !.

This analysis presupposes, however, that the condition q is in the *power* of the commanded person y, i.e. y may realise q or leave it unrealised as he wishes, since otherwise the alternative $\sim q$ is not real. A conditional imperative behaves differently, if y cannot or may not choose or decide the realisation of q:

If there is vodka in your glass, drink it!
If it rains tomorrow, use your umbrella!
If you see the enemy, raise the alarm!
Don't use the WC while the train is moving!

In these examples the act of commanding is empty or *pointless*, if the condition is already false at the moment when the command is uttered (e.g. the glass is empty). Likewise the command is cancelled or *annulled*, if the condition is not eventually realised (e.g. it does not rain tomorrow). On the other hand, if the condition sentence becomes true in a situation (e.g. the train is moving), then the conditional command is transformed into a *non*-conditional command.

Let α be a sentence which is not in the power of the commanded person. Then the above interpretation for hypothetical imperatives may be expressed in the following way. When α is temporally definite,

$$!(p/\alpha_t) = \begin{cases} !p, \text{ if } \alpha \text{ is true at time } t \\ \text{empty/annulled, if } \alpha \text{ is false at time } t. \end{cases} \tag{18}$$

When α is a generic condition, which may be true on several occasions,

$$!(p/\alpha) = \begin{cases} !p, \text{ when } \alpha \text{ is true} \\ \text{no effect, when } \alpha \text{ is false.} \end{cases} \tag{19}$$

According to (18) and (19), conditional imperatives are defined in analogy with *conditional bets*: the bet is called off or annulled, if the condition turns out to be false.[9] It follows from this suggestion that the conditional obligation generated by (18) and (19) can be expressed by

$$\begin{gathered} \alpha_t \Rightarrow Op \\ \alpha \Rightarrow Op, \end{gathered} \tag{20}$$

where $\Rightarrow$ is the intensional Lewis-type conditional ('if it were ..., then it would be ...').[10] Thus, a condition α, which is not in the power of the commanded person, is outside the scope of the obligation operator O.

[9] For the idea of conditional bets, see de Finetti [6, pp. 108–109]. For the suggestions (18) and (19), see Niiniluoto [13, 14]. Dummett [5, pp. 8–9] suggests that 'a conditional command where the antecedent is in the power of the person given the order ... is always like a bet on the material conditional' (this is our (17)), and he adds that 'where the antecedent is not in the person's power, it is indeed possible to regard the conditional command as analogous to the conditional bet' (this is our (18)). However, Dummett eventually rejects the latter proposal, since he thinks that 'the right to a reward' is not a consequence of obedience to a command. I don't find this a good reason to reject (18), since the key elements of that proposal are obedience, disobedience, and non-obedience to a command rather than the consequences (gains and losses) of such behaviour. Kenny [9, p. 65] notes that the conditional command 'If α, bring it about that p' is 'simply void' if α is not the case. Unlike (19), Kenny's proposed formalisation, i.e. $!(\alpha \supset$ seeing-to-it by you that $p)$, still contains the material implication.

[10] See Niiniluoto [13, 14]. The analysis (20) has been supported on other grounds e.g. by Peter Mott and Brian Chellas, cf. [4, p. 276]. The proposal '$\alpha \supset Op$' would be obviously mistaken, since it would make $O(p/\alpha)$ true whenever α is false. Instead the truth value of $\alpha \Rightarrow Op$, when α is false, depends on the truth of Op in those α-worlds which are most similar to the actual world.

5 Conditional Intentions

On the basis of the analogy between commands and intentions we may now suggest that there are genuinely *disjunctive intentions* $I(p \vee q)$, where both alternatives p and q are in my power:

I travel to Moscow by train or by plane!

This kind of expression of intention is appropriate in situations, where I have not yet decided which alternative I shall choose.[11] It may be expressed also in a conditional form

I travel to Moscow by train, if I don't take the plane!

The sentence

$$A(q \supset p) \tag{21}$$

thus describes such a *conditional* intention $A(p/q)$, where I may still choose between the alternatives $\sim q$ and $q\&p$.

What about conditional intentions $A(p/\alpha)$, where the condition α is not in the power of the relevant agent? Examples of such cases are illustrated by

If it rains tomorrow, I intend to stay at home.
If the enemy attacks, I intend to raise the alarm.

Sellars [17] and Castañeda [3] have debated, whether the condition here belongs to the scope of the intention-operator or not. The former alternative would correspond to proposal (21), i.e. $A(\alpha \supset p)$, the latter to formula

$$a \Rightarrow Ap \tag{22}$$

which is analogous to conditional obligation (20). According to (22), conditional intentions would be *conditional intendings*: if α were true, I should have the intention that p.

The proposal (22) is clearly inadequate, however, since it does not take into account the peculiar *epistemic* character of intentions—which also makes them unlike obligations. A conditional obligation $O(p/\alpha)$ is changed into non-conditional Op when the condition α becomes true, even if we failed to notice or to have any knowledge about this. Instead, a conditional intention $A(p/\alpha)$ does not produce Ap merely by the realisation of condition α, but I have to know this $(K\alpha)$ or at least have sufficient evidence for the belief $(B\alpha)$ that α (see Figure 2).

A similar observation can be made in the case, where I have already in advance certain belief that condition α will be true (see Figure 3). My intention $A(p/\alpha)$ is then not genuinely conditional, but may be expressed by

[11] This means that $I(p \vee q)$ does not guarantee Ip or Iq, and the corresponding formula

$$A(p \vee q) \supset (Ap \vee Aq)$$

is not valid.

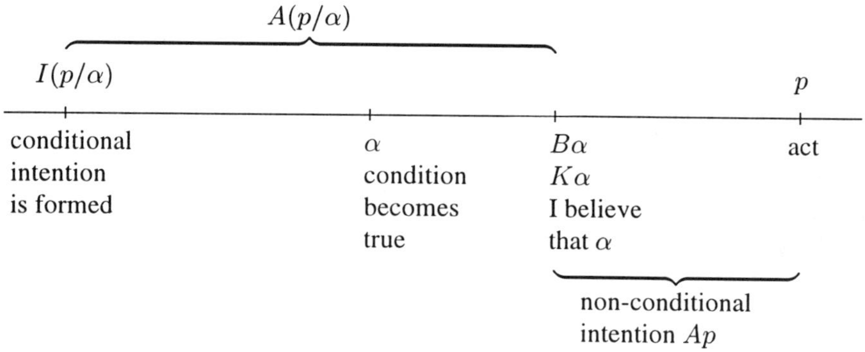

Figure 2:

I intend to do p, *when* α becomes true.

Sentence α is not anymore a condition for my intending, but rather a temporal *threshold* after which the time for doing p has arrived.

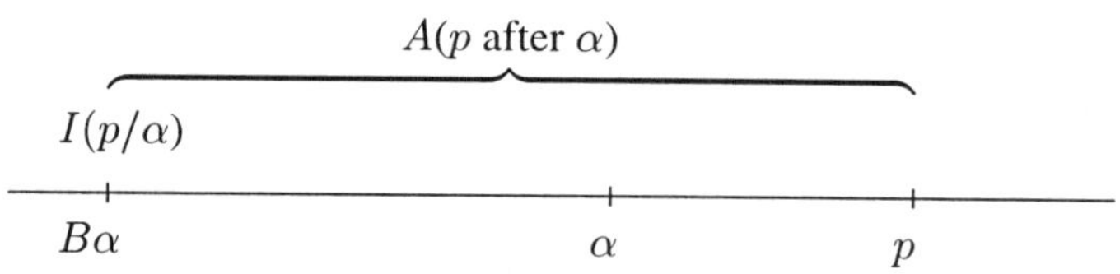

Figure 3:

Instead of (22), the logical form of the conditional intention $A_a(p/\alpha)$ generated by $I(p/\alpha)$, when condition α is not in the power of person a, is therefore

$$B_a a \Rightarrow A_a p, \tag{23}$$

i.e. if a believed that α, then a would have the non-conditional intention that p. Sentence (23) can be interpreted in a natural way to assert that a conditional intention exists and is preserved over time as a *mental disposition* in the mind of the intending agent.

As an example of the applications of formula (23), let us formalise the sentence

I intend to kiss Olga, if she gives me her permission. (24)

As the condition does not depend on my choice, this sentence may be written in the form

$$(\exists x)[B(x = \text{ Olga } \& \ x \text{ gives 2 permission}) \Rightarrow A(\text{I kiss } x)].$$

This formalisation explains why the pronoun 'she' occurs in a referentially opaque context in (24), even though it is outside the scope of the intention-operator A: 'she' belongs to the scope of the belief-operator B.

6 Detachment Rules

Proposal (22) for conditional intentions is compatible with the rule

$$\frac{\begin{array}{l} a \text{ intends to do } p, \text{ if } \alpha \\ a \text{ believes that } \alpha \end{array}}{a \text{ intends to do } p} \tag{25}$$

(cf. Kim [11, p. 261], Tuomela [20, p. 155], [21, p. 212]).[12] In other words, the rule

$$\frac{\begin{array}{l} A(p/\alpha) \\ B\alpha \end{array}}{Ap} \tag{26}$$

is valid, since

$$\frac{\begin{array}{l} B\alpha \Rightarrow Ap \\ B\alpha \end{array}}{Ap} \tag{27}$$

but the rule

$$\frac{\begin{array}{l} A(p/\alpha) \\ \alpha \end{array}}{Ap}$$

is not valid. It would correspond to the 'factual detachment' principle for conditional obligations,

$$\frac{\begin{array}{l} q \Rightarrow Op \\ q \end{array}}{Op} \tag{28}$$

(cf. formula (20)), but it fails for reasons that make (22) inadequate as an analysis of condition intentions.

The situation is different in an interesting way, if the condition of the intention is in my power (see formula (21)). Just as obligations satisfy a principle of 'deontic

[12] This inference should be distinguished from the so-called *practical syllogism* which combines a non-conditional intention with a conditional belief. See [23].

detachment'

$$\frac{\begin{array}{l} O(q \supset p) \\ Oq \end{array}}{Op,} \tag{29}$$

a temporally interpreted rule

$$\frac{\begin{array}{l} A(q \supset p) \\ Aq \end{array}}{Ap} \tag{30}$$

seems reasonable, since it is equivalent to

$$\frac{\begin{array}{l} A(\sim q \vee p) \\ Aq \end{array}}{Ap} \tag{31}$$

If my intention is to do $\sim q$ or p, and *afterwards* I form the intention to do q, then my disjunctive intention is transformed to Ap.

On the other hand, the rule

$$\frac{\begin{array}{l} A(q \supset p) \\ q \end{array}}{Ap}$$

is not generally valid or even meaningful. For example, if I have already done q, it is not reasonable for me to command myself disjunctively $\sim q \vee p!$, since $\sim q$ is not any more a possibility in my power. Further, if my intention is initially disjunctive $\sim q \vee p$, and the possibility $\sim q$ is eliminated or q is realised without my knowing it, a non-conditional intention is not produced. But if I (consciously) do q, then I cannot fail to detach an intention from a conditional one, so that the following principle appears to be acceptable:

$$\frac{\begin{array}{l} \text{I intend to } p, \text{ if } q \\ \text{I do } q \end{array}}{\text{I intend to } p} \tag{32}$$

University of Helsinki, Finland

References

[1] E. Anscombe. *Intention*, Blackwell, Oxford, 1957.

[2] H.-N. Castañeda. *Thinking and Doing*, D. Reidel, Dordrecht, 1975.

[3] H.-N. Castañeda. Reply to Wilfrid Sellars: intending and the logic of intentions. In [18], pp. 419–434, 1983.

[4] B. Chellas. *Modal Logic: an Introduction.* Cambridge University Press, Cambridge, 1980.

[5] M. Dummett. Truth. In *Truth and Other Enigmas*, pp. 1–24, Duckworth, London, 1978.

[6] B. de Finetti. Foresight: its logical laws, its subjective sources. In *Studies in Subjective probability*, H. E. Kyburg, Jr. and H. Smolker (eds), pp. 93–158. J. Wiley, New York, 1964.

[7] J. Hintikka. *The Intentions of Intentionality*. D. Reidel, Dordrecht, 1975.

[8] H.-P. Innala. *Is Deontic Logic Possible?*. PhD Thesis, Victoria University of Wellington, 1989.

[9] A. Kenny. Practical reasoning and rational appetite. In [15], pp. 63–80, 1978.

[10] A. Kenny. *Will, Freedom and Power*, Blackwell, Oxford, 1975.

[11] J. Kim. Intention and practical inference. In [12], pp. 249–269, 1976.

[12] J. Manninen and R. Tuomela, eds. *Essays on Explanation and Understanding*. D. Reidel, Dordrecht, 1976.

[13] I. Niiniluoto. Truth and legal norms. In *Conditions of Validity and Cognition in Modern Legal Thought*, N. MacCormick *et al.* eds, pp. 168–190. ARSP, Beiheft No. 25, Franz Steiner Wiesbaden, 1985.

[14] I. Niiniluoto. Hypothetical imperatives and conditional obligations. *Synthese*, **66**, 111–133, 1986.

[15] J. Raz, ed. *Practical Reason*, Oxford University Press, Oxford, 1978.

[16] A. Ross. *Directives and Norms*, Routledge and Kegan Paul, London, 1968.

[17] W. Sellars. Conditional promises and conditional intentions. In [18], pp. 195–222, 1983.

[18] J. Tomberlin, ed. *Agent Language and the Structure of the World*, Hackett, Indianapolis, 1983.

[19] J. Tomberlin, ed. *Hector–Neri Castañeda*. Profiles, Vol. 6, D. Reidel, Dordrecht, 1986.

[20] R. Tuomela. *Human Action and Its Explanation*, D. Reidel, Dordrecht, 1977.

[21] R. Tuomela. *A Theory of Social Action*, D. Reidel, Dordrecht, 1984.

[22] R. Tuomela and K. Miller. We-intentions. *Philosophical Studies*, **53**, 115–137, 1988.

[23] G. H. von Wright. *Practical Reason*, Blackwell, Oxford, 1983.

12

On Epistemic Modal Predicate Logic

M. N. Bezhanishvili

In their earlier investigations Frege and Russell have specially turned their attention to logical analysis of complex sentences with subordinate clauses, which are introduced by the connective *that* and the verbs *know, believe, doubt* and so on (e.g. 'Columbus believed that he had paved a new nautical way to India'). Following Russell and Ducass, relations, which are expressed by these complex sentences, are called propositional or epistemic attitudes. The truth of those sentences does not always depend on truth values of the subordinate clauses, it is not their function. Therefore, replacement of the subordinate clause by any of the equivalent sentences can change the truth value of the complex one. For a long time there was no satisfactory theory of logical analysis of these sentences. Unsuccessful was, ultimately, also the search of such criteria of identity of senses and synonymity of sentences, which guarantee permissibility of the above mentioned replacement.

Consideration of alternative states of affairs, situations, which are or are not compatible with knowledge, belief of a person, an agent of attitudes, is characteristic of contemporary, more convincing approaches. This means that verbs: know, believe, etc., are joined to modal notions of a special kind—to so-called epistemic (personal) modalities. Distinct approaches to explication and investigation epistemic attitudes are certainly connected with informal, intuitive treatment of their fundamental peculiarities.

In [1, 2] and [3] there was set out one of the possible approaches to semantical analysis of epistemic and doxatic propositional logic on the basis of the partial descriptions of possible worlds. The semantics, which was suggested in them, is a certain combination of usual possible worlds and Hintikka's model sets. It enables us to reproduce formally one of the main peculiarities of knowledge and belief of an agent of attitude—the limitation of his horizon, existence of gaps in his knowledge, narrowness of his opinion. Each of the systems considered there had contained one epistemic or doxatic modal operator (with a fixed agent of attitude), which informally means: 'someone knows (believes) that . . . '.

The aim of this paper is to generalize this approach for a predicate case. Though we consider only two epistemic systems each of them containing one epistemic operator with a fixed agent too. Other predicate systems can be formulated likewise.

The considered systems are capable of expressing only those epistemic attitudes

P. I. Bystrov and V. N. Sadovsky (eds.), Philosophical Logic and Logical Philosophy, 181–201.

in which the subordinate clause does not contain the name of a person occurring in the main sentence, in as far as the agent of attitude is beforehand fixed and the systems do not contain any individual constants. There is also no need to differentiate two kinds of individual variables, personal and impersonal, in them by the same cause (cf. [5]). Systems more rich from the epistemic view point in which the name of an agent can occur in subordinate clauses, will be considered in another paper. The problems connected with identity and description will also be considered there.

1 The Predicate systems EpT and Ep4

The alphabet of language contains an unlimited list of individual variables x, y, z and of n-adic predicate letters $P^n, Q^n, R^n (n > 0)$ with or without numerical subscripts (a 0-adic predicate letter is as usual called a propositional letter); the classical signs: the connectives $\neg, \&, \vee$ and the quantifiers $\forall, \exists$; one modal sign: the epistemic operator $\Box$, and also brackets. Other connectives are introduced by suitable definitions. The formulas are defined as usual (for example, simultaneously permitting free and bound occurrences of the same individual variable in a formula, and accepting corresponding convention: the occurrence of a variable in a formula is bounded by the most inside quantifier in the domain of which it occurs). We define usually an atomic formula, a modalized and a fully modalized one, the modal rank of occurrence of a subformula in any formula and the modal rank of a formula (see, e.g. [2]).

We supplement the list of axioms and rules of inference of [2] in order to formulate the epistemic first-order predicate systems EpT and Ep4. In particular, we add to the axiom group A all quantifier formulas of any formulation of the complete axioms of classical first-order predicate calculus; to the axiom group B:

1. $\Box\neg\neg P \equiv \Box P$,
2. $\Box(P\&Q) \equiv \Box P\&\Box Q$,
3. $\Box(P\&(Q \vee R)) \equiv \Box((P\&Q) \vee R)$,
4. $\Box P \supset \Box(P \vee Q)$,
5. $\Box Q \supset \Box(P \vee Q)$,
6. $\Box\neg(P\&Q) \equiv \Box(\neg P \vee \neg Q)$,
7. $\Box\neg(P \vee Q) \equiv \Box(\neg P\&\neg Q)$,

we add the following list of formulas:

8. $\Box\forall x P(x) \supset \Box P(y)$,
9. $\Box P(y) \supset \Box\exists x P(x)$,

10. $\Box\forall x\neg P(x) \equiv \Box\neg\exists x P(x)$,

11. $\Box\neg\forall x P(x) \equiv \Box\exists x\neg P(x)$,

12. $\Box\forall x(p(x) \vee Q) \supset \Box(\forall x P(x) \vee Q)$,

13. $\Box\exists x(P(x)\&Q) \supset \Box(\exists x P(x)\&Q)$.

Further, we correspondingly supplement the rest of the rules of inference of the classical first-order predicate calculus (their usual names for references to them will be used later) to the rules of inference of classical propositional calculus (namely to R1.1, rule of substitution for propositional letters, and R1.2, rule of modus ponens, of [2]). Finally, to the modal rules of inference of [2]:

R2. From $\Box A \supset \Box C$ and $\Box B \supset \Box C$, if A, B and C are nonmodalized, to infer $\Box(A \vee B) \supset \Box C$,

R3. From A, if it is fully modalized, to infer $\Box A$.

We add two following rules:

R4. From $\Box B \supset \Box A(x)$ to infer $\Box B \supset \Box\forall x A(x)$,

R5. From $\Box A(x) \supset \Box B$ to infer $\Box\exists x A(x) \supset \Box B$, where b does not contain free occurrences of x (in both cases).

The epistemic first-order predicate system EpT contains, except for the above mentioned axioms and the rules of inference, the axioms

1. $\Box P \supset P$ and
2. $\Box(\Box P \supset P)$ of axiom group C of [2]. Ep4 is obtained from EpT by adding of the axiom
3. $\Box P \supset \Box\Box P$ of axiom group C of [2].

In the propositional parts of those epistemic and doxatic systems the rules of replacement of the equivalent formulas have been restricted (see [3, 2]). Let the modal rank of the occurrence of A in a provable formula C be zero. Then the replacement of A by B in C gives again a provable formula as usual, if $A \equiv B$ is also provable. But if the modal rank of the occurrence of A in a provable formula C is more than 0, then the replacement of A by B in C gives again a provable formula, if $A \equiv B$ is provable and besides A and B are fully modalized or A and B are nonmodalized by the formulas $\Box A = \Box B$ and $\Box\neg A = \Box\neg B$ are provable.

The proofs of these principles were carried out in [3] (see Theorem 1) by means of the following auxiliary statements:

a. Let A and B be nonmodalized, and let the modal rank of occurrence of A in $C(A)$ be 0. Then, if $\Box A \equiv \Box B$ and $\Box\neg A \equiv \Box\neg B$ are provable, then $\Box C(A) \equiv \Box C(B)$ and $\Box\neg C(A) \equiv \Box\neg C(B)$ are provable too ([3, Lemma 1]).

b. If A and B are fully modalized and $A \equiv B$ is provable, then $\Box C(A) \equiv \Box C(B)$ is provable too ([3, Lemma 2]).

c. If A and B are nonmodalized and formulas $\Box A \equiv \Box B, \Box\neg A \equiv \Box\neg B$ are provable, then $\Box C(A) \equiv \Box C(B)$ is provable too ([3, Lemma 3]).

These statements and hence the rules of replacement of equivalent formulas with the same restrictions held also for the first-order predicate case.

The statement (A) is proved by induction with respect to the depth d of occurrence of A in C (instead of $C(A)$ we shall write also C_A to mark a replacing subformula A of C, and C_B will denote the result of replacement of A by B in C). When $d > 0$, two additional cases must be considered for the epistemic first-order predicate systems: C_A has a form $\forall x D_A(x)$ or $\exists x D_A(x)$. According to the inductive hypothesis $\Box D_A(x) \equiv \Box D_B(x)$ and $\Box\neg D_A(x) \equiv \Box\neg D_B(x)$ are provable.

Let C_A have a form $\forall x D_A(x)$. We must prove

a. $\Box\forall x D_A(x) \equiv \Box\forall x D_B(x)$ and

b. $\Box\neg\forall x D_A(x) \equiv \Box\neg\forall x D_B(x)$.

a.

$$\begin{aligned} \Box\forall x D_A(x) \supset \Box D_A(x) & \quad - \quad \text{by substitution in axiom B8,} \\ \supset \Box D_B(x) & \quad - \quad \text{by the inductive hypothesis,} \\ \supset \Box\forall x D_B(x) & \quad - \quad \text{by R4 (the condition holds).} \end{aligned}$$

Likewise,

$$\begin{aligned} \Box\forall x D_B(x) \supset \Box D_A(x) & \quad - \quad \text{by substitution in axiom B8,} \\ \supset \Box D_A(x) & \quad - \quad \text{by the inductive hypothesis,} \\ \supset \Box\forall x D_A(x) & \quad - \quad \text{by R4 (the condition holds),} \end{aligned}$$

and therefore $\Box C_A \equiv \Box C_B$.

b.

$$\begin{aligned} \Box\neg D_A(x) \supset \Box\neg D_B(x) & \quad - \quad \text{by the inductive hypothesis,} \\ \supset \Box\neg\forall x D_B(x) & \quad - \quad \text{by substitution in B9 (and by B11),} \end{aligned}$$

whence it follows $\Box\neg\forall x D_A(x) \supset \Box\neg\forall x D_B(x)$ according to the rule of inference R5 (the condition holds). Likewise,

$$\begin{aligned} \Box\neg D_B(x) \supset \Box\neg D_A(x) & \quad - \quad \text{by the inductive hypothesis,} \\ \supset \Box\neg\forall x D_A(x) & \quad - \quad \text{by substitution in B9 (and by B11).} \end{aligned}$$

whence we obtain $\Box\neg\forall x D_B(x) \supset \Box\neg\forall x D_A(x)$ again according to R5, and therefore $\Box\neg C_A \equiv \Box\neg C_B$.

The case, when C_A has a form $\exists x D_A(x)$, is considered likewise.

The statement (a) is proved.

As in [3], we shall call quasiatoms of A all such atomic formulas of A and all its such subformulas of the form $\Box B$, the modal rank of occurrences of which in A is 0 (as quasiatoms of A we can choose also all such atomic formulas and fully modalized subformulas of A, by which we can form A without the formation rule for $\Box$). Prenex normal formulas are defined as usual, and prenex normal formulas formed by quasiatoms shall be called such formulas, in which all those quantifiers that do not occur in any quasiatom are initially placed. It is easy to establish the following consequence of the statement (a): for every formula C there is a prenex normal formula GD formed by quasiatoms of C such that $\Box C \equiv \Box GD$ is provable (where G is quantifier prefix obtained by passage all those quantifiers of C which do not occur in quasiatoms of C, and D is a formula constructed by quasiatoms of C only by means of classical signs).

The consequence of the statement (a) can be stated in the following way: for each quasiatom of C in which we have n free occurrences of individual variables $x_1, \ldots, x_n$ (not necessarily all different), we choose n-adic predicate letter P^n, which does not occur in C, and each of these quasiatoms is replaced in C by the atomic formula $P^n(x_1, \ldots, x_n)$. In result we obtain nonmodalized formula C'. On the other hand, for each equivalence $A \equiv B$, which is used by transforming any nonmodalized formula into usual prenex form, in our systems there are proved formulas $\Box A \equiv \Box B$ and $\Box\neg A = \Box\neg B$. Therefore, according to the statement (a), formula $\Box C' \equiv \Box GD'$ is provable too, whence by means of suitable substitutions we can easily obtain $\Box C \equiv \Box GD$.

The statement (b) is proved by induction with respect to the modal rank r of occurrence of A in C_A. If $r = 0$, then either A coincides with C_A, and we easily infer $\Box C_A \equiv \Box C_B$, so far as A and B are fully modalized, or A is in C_A only in the scope of classical signs. Hence we can choose A as a quasiatom of C_A. According to the consequence of the statement (a), $\Box C_A = \Box GD_A$ is provable. Consider the formula D_A without the quantifier prefix G. By the Lemma 2 [3] (which was already established in [3] for the propositional case), formula $\Box D_A \equiv \Box D_B$ is provable, whence by means of suitable binding of variables we infer $\Box GD_A \equiv \Box GD_B$ (this is carried out by the axioms B8, B9 and by the rules of substitution R4 and R5).

Therefore,

$$\begin{aligned} \Box C_A &\equiv *\Box GD_A, \\ &\equiv \Box GD_B, \\ &\equiv \Box C_B. \end{aligned}$$

The case, when $r > 0$, may be considered as in the proof of [3, Lemma 2].

The statement (c) is proved by the induction with respect to the modal rank of the occurrence of A in C_A and by auxiliary statements (a) and (b) as in [3]. Finally, from the statements (a), (b) and (c), it follows as in [3] that the above formulated rules of replacement of equivalent formulas hold for predicate case (cf. Theorem 1

[2, 3]). They permit us, in particular, to eliminate from every formula the signs & or $\vee$ and $\forall$ or $\exists$, as far as the suitable equivalences are provable in our systems.

Later on EpT and Ep4 will simply denote the corresponding epistemic first- order predicate systems, if contrary is not reserved.

Semantics of the epistemic and doxatic propositional systems in [1, 2] and [3] was in fact described in terms of Hintikka's model sets. In the present paper more widely-spread formulations will be used.

EpT-model structure is an ordered quadruple $[H, W, R, D]$ where H is a set, W is any non-empty subset of H, $R \subseteq H \times H$ and R is reflexive on H, D is a domain function, defined on H, such that $D(v)$ is a non-empty set for all $v \in H$, and

d. If $(v, w) \in R$, then $D(v) \subseteq D(w), \quad v, w \in H.$

If besides R is a transitive on H, the EpT -model structure will be called Ep4-model structure.

Let $\mathcal{F}$ be the set of all predicate letters P^n of our language ($n \geq 0$). EpT-, Ep4-model M is a pair $[S, E]$ where S is EpT-, Ep4- model structure and E is a binary partial (evaluation) function defined on $\mathcal{F} \times H$ such that if $n = 0, E(P^n, v) = T$ or F or neither T nor F. In the first two cases we shall say that E is defined for P^n, v and write $!E(P^m, v)$; in the third case we shall say that E is undefined for P^n, v and write non $!E(P^n, v)$. Further, if $n > 0, E(P^n, v)$ is a partial function from $[D, (v)]^n$ to $\{T, F\}$ where $[D(v)]^n$ is the m-times Cartesian product of the set $D(v)$ on itself.

Remark 1.1 Naturally a question arises: how should it be motivated that we permit cases of undefined values of propositional and predicate letters from the epistemic point of view? Consider an example. Suppose $G(n)$ is the informal property: there is a pair of prime numbers k and l such that n is a natural number and $2n = k + 1$. Further, let $B(n)$ mean: for every natural number $m \geq nG(m)$. It is obvious that when $n = 1$, then $B(n)$ is false. But when $n > 1$, establishment of truth or falsity of $B(n)$ every time depends on the solution of the problem: has every natural number, more than 1, the property G? So far nobody knows the answer to the latter question, though probably many believe that $G(m)$ is true for all such m that $m > 1$. But this is only belief and is not a knowledge, because we have no method either for stating $B(n)$, when $n > 1$, or for constructing counterexample. So, from the epistemic viewpoint this is a gap, a white spot in our knowledge. But classically, i.e. for the ontological approach (from God's position), this property B is quite defined (a philosopher would express this situation by words: when $n > 1, B(n)$ is defined *as such*, but is undefined for our knowledge). And there is the watershed between the classical (ontological) and the epistemic (cognitive) viewpoints. For the epistemic approach, every statement, which supposes a solution to any open problem, is undefined, as well as every senseless statement.

Let, further,

$$U = \bigcup_{v \in H} D(v).$$

Given an EpT- or Ep4-model M, we can find for each formula A of the system EpT or Ep4 a value T or F for each $v \in H$ relative to a fixed assignment of elements of U to free individual variables of A, if $E(A, v)$ is defined.

If A is an atomic formula, it is either a propositional letter P^0 or it is a formula of the form $P^n(x_1, \ldots, x_n), n \geq 1$. In the first case $E(P^0, v)$ is already given by the model. In the latter case, suppose that elements $\alpha_1, \ldots, \alpha_n$ of U are assigned to $x_1, \ldots, x_n$, and let $E(P^n, v)$ be a partial function L; then relative to this assignment, if $v \in W, E(P^n(x_1, \ldots, x_n), v) = T$ iff $L(\alpha_1, \ldots, \alpha_n) = T$; otherwise $E(P^n(x_1, \ldots, x_n), v) = F$. And if $v \in H \backslash W, E(P^n(x_1, \ldots, x_n), v) = T$ iff $L(\alpha_1, \alpha_n) = T; E(P^n(x_1, \ldots, x_n), v) = F$ iff $L(\alpha_1, \ldots, \alpha_n) = F$; otherwise non $!E(P^n(x_1, \ldots, x_n), v)$.

Given this assignment to atomic formulas, we can build up the assignment to complex formulas by induction. The conditions of evaluation of formulas of the form $\neg A, A\&B, A \vee B$ for any v from H by the fixed assignment of elements of U to all free individual variables of A and B are the same as in [2]. They coincide with Łukasiewicz's three-valued tableaux (for negation, conjunction and disjunction) but instead of the third undefined value we have a gap (absence of value).

Conditions to quantifiers.

Let $x_1, \ldots, x_n$ be all distinct free individual variables of a formula $\forall y A(x_1, \ldots, x_n, y)$, respectively $\exists y A(x_1, \ldots, x_n, y)$, and the elements $\alpha_1, \ldots, \alpha_n$ of U are correspondingly assigned to them. We say that $E(\forall y A(x_1, \ldots, x_n, y), v) = T$ for the given assignment iff $D(A(x_1, \ldots, x_n, y), v) = T$, when for the same assignment $\alpha_1, \ldots, \alpha_n$ to $x_1, \ldots, x_n$ we assign to y any $b \in D(v)$. And $E(\forall y A(x_1, \ldots, x_n, y), v) = F$ for the given assignment iff there is $bD(v)$, such that $E(A(x_1, \ldots, x_n, y), v) = F$, when for the assignment $\alpha_1, \ldots, \alpha_n$ to $x_1, \ldots, x_n b$ is assigned to y; otherwise non $!E(\forall y A(x_1, \ldots, x_n, y), v)$.

Analogously, we say that $E(\exists y A(x_1, \ldots, x_n, y), v) = T$ for the given assignment iff there is $b \in D(v)$ such that $E(A(x_1, \ldots, x_n, y), v) = T$, when for the assignment $\alpha_1, \ldots, \alpha_n$ to $x_1, \ldots, x_n$ b is assigned to y; $E(\exists y A(x_1, \ldots, x_n, y), v) = F$ for the given assignment iff $E(A(x_1, \ldots, x_n, y), v) = F$, when for the same assignment $\alpha_1, \ldots, \alpha_n$ to $x_1, \ldots, x_n$ we assign to y any $b \in D(v)$; otherwise non $!E(\exists y A(x_1, \ldots, x_n, y), v)$.

Finally, as in the propositional case (see e.g. [2]), for a given assignment of elements of U to all free individual variables of $AE(\Box A, v) = T$ iff for the same assignment $E(A, w) = T$ (and therefore $E(A, w)$ is defined) for every w from H, such that $(v, w) \in R$; otherwise $E(\Box A, v) = F$ (namely iff for the same assignment $E(A, w) = F$ or non $!E(A, w)$ for some w from H such that $(v, w) \in R$).

It is obvious that for a given assignment $E(A, w)$ is always defined, when $W \in W$ or A is fully modalized.

If $H \backslash W = \varnothing$, then EpT-, Ep4-model is T-, S4- model.

Models for other epistemic and doxatic first-order predicate systems, whose propositional parts were considered in [1, 2] and [3], are defined analogously.

A is true in a model $M = [H, W, R, D, E]$ for v from H by a given assignment of elements of U to all free individual variables of A if $E(A, v) = T$ for the same assignment. A is true in a model M for a given assignment of elements of U to all free individual variables of A, if for the same assignment A is true for every w from W. A is valid in M, if A is true in M for every assignment of elements of U to al free individual variables of A. And, finally, A is valid, if it is valid in every model.

Consistency of the propositional parts of EpT and Ep4 was proved in [2]. Further, we can easily establish validity of the axioms B8–B13 by immediate verification. The proofs, that the rules of substitution for predicate letter and free individual variables, and also the rules of renaming of the bounded individual variables and introduction of quantifiers preserve validity in EpT and Ep4, are usual. It remains to show that rules R4 and R5 preserve validity too.

Let $y_1, \ldots, y_n$ be distinct individual variables among which are all different from x free individual variables occurring in A and $B(x)$.

R4. Suppose that the rule of inference R4 does not preserve validity. So, let $\Box A \supset \Box B(x)$ be valid and $\Box A \supset \Box \forall x B(x)$ be invalid. Then there is a model $M = [H, W, R, D, E]$ which refutes the latter formula. Thus, for some $w \in W E(\Box A \supset \Box \forall x B(x), w) = F$ for corresponding assignment of elements $\alpha_1, \ldots, \alpha_n$ of U to the variables $y_1, \ldots, y_n$, and, therefore, $E(\Box A, w) = T$ and $E(\Box \forall x B(x), w) = F$ for the same assignment. But then for some $v \in H$, such that $(w, v) \in R$ (in EpT-modle R is reflexive and in Ep4-model R is reflexive and transitive) $E(\forall x B(x), v) = F$ or non $!E(\forall x B(x), v)$ for the same assignment. This means that there is an element b from $D(v)$ for which $E(B(x), v) = F$ or non $!E(B(x), v)$, when to all different from x free individual variables of $B(x)$ the above mentioned elements of U are correspondingly assigned and to the variable x the element b is assigned. Hence, for the assignment $\alpha_1, \ldots, \alpha_n, b$ to the variables $y_1, \ldots, y_n, x$ respectively, $E(\Box B(x), w) = F$ and, as x does not occur free in A, $E(\Box A, w) = T$ for the same assignment. Therefore, $E(\Box A \supset \Box B(x), w) = F$, when the elements $\alpha_1, \ldots, \alpha_n$ of U are assigned to the free individual variables $y_1, \ldots, y_n$, respectively, and the element b from $D(v)$ is assigned to the variable $x (D(v) \subseteq U)$. However, this contradicts our assumption on validity of the premise of the rule of inference R4.

R5. Analogously, it is stated that the rule of inference R5 preserves validity too. In fact, if we suppose again contradictory, as in the previous case, then there must be such a model M, refuting the formula $\Box \exists x B(x) \supset \Box A$, that for some $w \in W E(\Box \exists x B(x), w) = T$ and $E(\Box A, w) = F$ for the corresponding assignment of the elements $\alpha_1, \ldots, \alpha_n$ of U to the variables $y_1, \ldots, y_n$. As A does not contain free occurrences of x, $E(\Box A, w) = F$ also for any element b of U for the assignment $\alpha_1, \ldots, \alpha_n, b$ to the variables $y_1, \ldots, y_n, x$, respectively. According to our supposition, $\Box B(x) \supset \Box A$ is valid and hence $E(\Box B(x), w) = F$ for the same assignment (for any b). But then there is an element v_0 from H such that $(w, v_0) \in R$ and $E(B(x), v_0) = F$ or non $!E(B(x), v_)) $ for the same assignment. On the other hand, according to our supposition, $E(\Box\exists x B(x), w) = -T$ for the assignment $\alpha_1, \ldots, \alpha_n$ to the variables $y_1, \ldots, y_n$ and, therefore, for every v, such that $(w, v) \in R$, in particular for v_0 $E(\exists x B(x), v_0) = T$ and hence there is an element b_0 from $E(v_0) \subseteq U$, such that $E(B(x), v_0) = T$, when $\alpha_1, \ldots, \alpha_n$ are correspondingly assigned to $y_1, \ldots, y_n$ and b_0 is assigned to x. But as we have already stated, proceeding from our assumption, $E(B(x), v_0) = F$ or non $!E(B(x), v_0)$ for the assignment $\alpha_1, \ldots, \alpha_n, b$ to $y_1, \ldots, y_n, x$ by the same $\alpha_1, \ldots, \alpha_n$ and arbitrary b from U, and when we choose b_0 instead of b, we obtain the contradiction.

Thus, we have stated that all the axioms of EpT and Ep4 are valid and the rules of inference preserve validity. Hence, if $\vdash A$ in EpT, Ep4, then $\models A$ in EpT, Ep4 (see [2, Theorem 2]).

1.1 Semantic Tableaux. Completeness

In [1] the modified Kripke style the method of construction of semantic diagrams for the propositional parts of EpT and Ep4 was described, and in [1] and [2] the completeness of these systems was proved. The rules for the construction of semantic tableaux in [1] were formulated for the signs $\neg$, $\vee$ and $\Box$ (as & can be eliminated from each formula). For the predicate case it is sufficient to add to those rules the rule for universal quantifier below (as one of the signs $\forall$ and $\exists$ can also be eliminated from each formula). Notice that the upper line over formulas mean, as in [1] and [2], the absence of these formulas in the tableau (on the necessity of such information in the epistemic diagrams see [1]); as earlier, $\mathcal{R}$ is a binary relation between tableaux, which corresponds to the relation of accessibility between elements of H.

Π. $\overline{N\Pi}$. If a formula $\forall x A(x)$, correspondingly an expression $\overline{\neg\forall x A(x)}$, appears in a tableau t, then for each free individual variable y of every tableau t, then for each free individual variable y of every tableau t', such that $(t, t') \in \mathcal{R}$ put the formula $A(y)$, correspondingly the expression $\overline{\neg A(y)}$, in the tableau t.[1]

$N\Pi, \overline{\Pi}$. If a formula $\neg\forall x A(x)$, correspondingly an expression $\overline{\forall x A(x)}$, appears in a tableau t, then put in t the formula $\neg A(y)$, correspondingly the expression $\overline{A(y)}$, where y is a new individual variable, which did not occur earlier in any tableau.

To these quantifier rules we add one more rule: if in a tableau there appears no free individual variable and none of them is introduced by rules $N\Pi$, $\overline{\Pi}$, then for guaranteeing of applicability of the rules $\Pi, \overline{N\Pi}$ we put in the tableau a new free individual variable.

As in [1] and [2] we define an auxiliary, an alternative, an explicitly inconsistent, an inconsistent, a properly inconsistent, a consistent tableaux and a stage in

[1] We assume that if y is not free for substitution instead of x in $A(x)$, then we carry out corresponding renaming of the bound individual variables in $A(x)$.

constructing of a tableau.

Different from the propositional case, a consistent tableau of a diagram can contain an infinite set of formulas, which spends the all unlimited reserve of individual variables of our alphabet (for example, if in a consistent tableau t appears only one free individual variable x and t contains a formula of the form $\forall x \neg \forall y A(x,y)$, then the rule Π is applicable to them). According to this rule we put $\neg \forall y A(x,y)$ in t. To the later formula the rule $N\Pi$ is applicable and in accordance with them we put $\neg A(x,y)$ in t with new free variable y. But, as in t the new free variable y appears, the rule Π is again applicable to $\forall x \neg \forall y A(x,y)$ and we put $\neg \forall z A(y,z)$ in t with the renaming bound variable, as y is not free for the sub situation instead of x in $\neg \forall y A(x,y)$. To that formula the rule $N\Pi$ is applicable too and we put in $t \neg A(y,z)$ with new free variable z. So far as this process of alternate application of the rules Π and $N\Pi$ does not break off, in t it will be also included $\neg \forall y A(z,y), \neg A(z,x_1), \neg \forall y A(x_1,y), \neg A(x_1,y_1)$ and so on).

On the other hand, the aim of constructing of a diagram for A is finding a counter model for it, if A is refutable (or establishing of a certain property of its irrefutability, otherwise). Besides, we intend to construct a counter model M for A with such account that every element v of H, which corresponds to t, $D(v)$ will coincide with the set of all free individual variables occurring in t. However, if all reserve of individual variables will be already spent in t, we cannot always guarantee observance of the condition (d) of domain function in its general formulation. Hence we must spend economically the reserve of individual variables, although it is unlimited. For this purpose to each tableau t we may correspond unlimited set Y_t of individual variables by the condition that the sets, which correspond to distinct tableaux will be disjointed, and that by use of the rules $N\Pi$ and $\overline{\Pi}$ to formulas of t the new variables always appear in t from Y_t.

As in [1] and [2] we define usually the characteristic formula of a stage of a tableau and ancestors and descendants of a given formula in a tableau.

Let a formula B appear in a tableau t at the lth stage and all descendants of B appear in t by the use of one of the rules NN, ND, Π and $N\Pi$. We place the sign $\Box$ from the left of the formula B and of each of its descendant. Characteristic formulas of each stage m of thus transformed tableau, beginning with lth stage ($m \geq l$), shall be called quasi-characteristic formulas of the mth stage of a tableau t.

Completeness of the propositional part of EpT in [1] was stated by means of Lemmas 1–6 (more compact formulations of which are given in [2]), besides, Lemmas 1–5 were used to prove the statement: if a diagram for A is inconsistent (abbreviated, $\Vdash A$), then $\vdash A$ in EpT. Each of these Lemmas, except for the specific for EpT Lemma 4, holds for the propositional part Ep4 too, for which this statement is established by means of the specific for Ep4 Lemma 9 of [2] and Lemmas 1–3, 5 [1]. On the first-order predicate systems EpT and Ep4 the additional cases, connected with the use of rules for quantifiers Π, $\overline{N\Pi}$, $N\Pi$, $\overline{\Pi}$, arise only by generalisation of Lemmas 1 and 2 of [1] (and also by generalisation Lemma 6 of [1], in case EpT, and Lemma 10 of [2], in case Ep4, to prove the statement: if $\models A$, then $\Vdash A$).

When at the $(n+1)$th stage we use one of the rules for quantifiers, the fact that the items (a)–(c) of generalized Lemma 1 of [1]holds, may be proved by means of the theorems and derived rules of inference below (in a like manner as in the propositional case).

To establish (a) we use the theorem

$$(\neg\forall x A(x) \vee \neg A(y)) \supset \neg\forall x A(x)$$

and the derived rule:

from $\neg\neg\forall x A(x) \vee \neg\neg A(y)$ to infer $\neg\neg\forall x A(x)$,

where y does not occur free in $\forall x A(x)$. This condition will be held in a tableau which contains a formula of the form $\neg\forall x A(x)$ and when we apply the rule of inference $N\Pi$ to it, we put in the tableau $\neg A(y)$ with a new individual variable y. In the characteristic formula of the corresponding stage, this formula is presented by the disjunctive member $\neg\neg A(y)$ with the same y and to the formula $\neg\forall x A(x)$ there corresponds the disjunctive member $\neg\neg\forall x A(x)$.

To prove (b) we use the theorem

$$\Box(\neg\forall x A(x) v \neg A(y)) \supset \Box\neg\forall x A(x)$$

and the derived rule of inference:

from $\Box(\neg\neg\forall x A(x) \vee \neg\neg A(y)$ to infer $\Box\neg\neg\forall x A(x)$,

where y does not occur free in $\forall x A(x)$. This condition will be held by applying the rules $N\Pi$, $\overline{\Pi}$.

And to state (c) we respectively should use the theorems:

$$(\Box(\forall x A(x) \& A(y)) \supset \Box B) \supset (\Box\forall x A(x) \supset \Box B),$$
$$(\Box B \supset \Box(\neg\forall x A(x) \vee \neg A(y))) \supset (\Box B \supset \Box\neg\forall x A(x))$$

and the derived rules of inference:

from $\Box(\neg\forall x A(x) \& \neg A(y)) \supset \Box B$ to infer $\Box\neg\forall x A(x) \supset \Box B$,
$box from \Box B \supset \Box(\forall x A(x) \vee A(y))$ to infer $\Box B \supset \Box\forall x A(x)$,

where y does not occur free in B (in both cases). This condition will be held by any application of rules $N\Pi, \overline{\Pi}$.

So, Lemma 1 of [1] is established for the predicate case.

Lemma 2 in [1] was proved by Lemma 2.1 [1]. But the latter is not generalized directly for the predicate case. Hence, we first prove another auxiliary statement—Lemma 1.2 and by means of it Lemma 2 of [1] will be stated for the predicate case.

Lemma 1.2 *Let $\chi_{m+1}(t)$ and $\chi_m(t)$ are quasi-characteristic formulas of the $(m+1)$th and mth stages of a tableau $t(m > 1)$. Then if $\Box\chi_{m+1}(t)$ is provable in EpT, $\Box\chi_m(t)$ is provable too.*

PROOF. The proof is carried out quite similarly as for the item (c) of Lemma 1 [1], by means of the following theorems:

$$\Box(\neg\Box\neg\neg A \vee \neg A(\supset \Box\neg\Box\neg\neg A,$$
$$\Box(\neg\Box\neg(A \vee B) \vee \neg\Box\neg A \vee \neg\Box\neg B) \supset \Box\neg\Box\neg(A \vee B),$$
$$\Box(\neg\Box\forall x A(x) \vee \neg\Box A(y)) \supset \Box\neg\Box\forall x A(x)$$

and the derived rule:

$$\text{from } \Box(\neg\Box\neg\forall x A(x) \vee \neg\Box\neg A(uy)) \text{ to infer } \Box\neg\Box\neg\forall x A(x),$$

where y does not occur free in $\forall x A(x)$.

Now we may show that Lemma 2 [1] will be held in the predicate case to. The former proof of the item (a) of the Lemma holds (see [1]). Let us prove the item (b): if an explicitly inconsistent tableau t is auxiliary, then $\chi(t)^{\Box}$ is provable in EpT (Ep4).

According to the condition of the Lemma, the tableau t is explicitly inconsistent. Hence, the construction of t is finished as soon as formulas E and $\neg E$ or a formula E and an expression $\overline{E}$ appear in it.

Besides, t is auxiliary and, hence, at the initial stage $\overline{C}$ and $B_1, \ldots, B_k (k \geq 0)$ are put in it. If t contains E and $\neg E$, then both formulas can be descendants of $B_1, \ldots, B_k$ or descendants of $\overline{C}$ or one of them is a descendant of $B_1, \ldots, B_k$ and the other is a descendant of $\overline{C}$. But if t contains E and $\overline{E}$, then only the second or third from that possibility may hold for them. According to this, we should consider separately each of these three possibilities.

The first and the third ones are considered likewise as in [1]. It remains to consider the second possibility: let E and $\neg E$ or E and $\overline{E}$ be descendants of $\overline{C}$. E and $\neg E$ without upper lines may appear in t only after using the rules $\overline{K}$ or $\overline{NK}$ to the expression $\overline{C}$ or to one of its descendants. Therefore, t contains an expression of the form $\overline{\Box D}$ or $\overline{\neg\Box D}$, and E and $\neg E$ are their descendants. Suppose, without limiting our considerations, that E is a descendant of one of these expressions.

If t contains an expression $\overline{\Box D}$, then the rule $\overline{K}$ is applicable to the latter and in accordance with it formula $\neg\Box D$ is put in t. E may coincide with it, but cannot be its descendant, because to $\neg\Box D$ is applicable only rule NK, pointing out to start a new tableau. However, this is impossible as t is explicitly inconsistent.

If t contains an expression $\overline{\neg\Box D}$, then the rule $\overline{NK}$ is applicable to the latter and formula $\neg\neg\Box D$ is put in t. E either coincides with it or is its descendant, which is placed in t according to the rules NN, D, K, Π or $N\Pi$. In fact, the rules D and NK cannot be applied in this case, as an explicitly inconsistent tableau does not have alternative and auxiliary ones. But the rules $\overline{NN}, \overline{S}, \overline{ND}, \overline{K}, \overline{NK}, \overline{\Pi}$ and $\overline{N\Pi}$ cannot be applied, as $\neg\neg\Box D$ has not an upper line.

Recall that, if we ignore the order of disjunctive members, characteristic formula $\chi(T)$ has the form:

$$F \& B_1 \& \ldots \& B_k \supset \neg\neg C \vee G,$$

where F, G may be empty expressions, and $\chi(t)^{\Box}$ is

$$\Box(f\&B_1\&\ldots\&B_k) \supset \Box(\neg\neg C \vee G).$$

When $k = 0, \chi(t)^{\Box}$ coincides with $\Box\chi(t)$.

Suppose at first, as in the propositional case, that $k = 0$. So, E has the form $\neg\Box D, \neg\neg\Box D$ or it coincides with the immediate descendant $\Box D$ of the formula $\neg\neg\Box D$, which is placed into t using the rule NN, or it is a descendant of $\Box D$. In first three cases the item (b) of Lemma 2 proves as in [1]. Consider the latter case: E is a descendant of $\Box D$. Let among the ancestors of E and the descendants of $\Box D$ the formula $\Box B$ be the last appeared in t such a formula, which begins with the sign $\Box$. To $\Box B$ we may apply the rule K. Let in consequence of using that rule in t be placed formula B at the $(m+1)$th stage. Then, if we reconstruct properly the tableau t, placing the sign $\Box$ in the left side of B and all its descendants, then the quasi-characteristic formula of $(m+1)$th stage will coincide with the characteristic formula of mth stage. Now according to Lemma 1.2, for every stage $l(l > m)$, if $\Box\chi_{l+1}(t)$ is provable in EpT, then $\Box\chi_1(t)$ is provable too. Therefore, it is sufficient to prove that $\Box\chi_n(t)$ is provable in EpT, where n is the last stage in the construction of t.

If t contains E and $\neg E$, then for $\neg E$ we also have two possibilities; $\neg E$ is a descendant of B or not. But if t contains E and $\overline{E}$, then, because of the upper line, $\overline{E}$ cannot be a descendant of B. In accordance with that, when $\neg E$ is a descendant of B, then $\chi_n(t)$, ignore the order of disjunctive members, has the form

$$\Box(\neg\Box E \vee \neg\Box\neg E \vee G),$$

and when $\neg E$, then $\chi_n(t)$ is $\Box(\neg\Box E \vee \neg\neg E \vee G)$.

Let us show, that in both cases $\Box\chi_n(t)$ is provable in EpT (and therefore, in Ep4). Actually, it is easy to verify that $\Box\neg\Box\neg(\neg E \vee \neg\neg E)$ is provable in EpT. Further we have

$$\begin{aligned}\Box\neg\Box(\neg E \vee \neg\neg E) \quad &\supset \Box\neg\Box(\neg\neg E\&\neg\neg\neg E),\\ &\supset \Box\neg(\Box\neg\neg E\&\Box\neg\neg\neg E),\\ &\supset \Box(\neg\Box E \vee \neg\Box\neg E),\\ &\supset \Box(\neg\Box E \vee \neg\Box\neg E \vee G),\end{aligned}$$

and, hence, $\Box\chi_n(t)$ is provable in EpT. But then according to Lemma 1.2, in EpT is provable also $\Box\chi_{m+1}(t)$, which coincides with $\Box\chi_m(t)$ and by the Lemma 1 (c) [1], for any stage less than m, in EpT is provable $\Box\chi(t)$, i.e. $\chi(t)^{\Box}$.

In the second case we have

$$\begin{aligned}\Box()\neg\Box E \vee E) \quad &\supset \Box(\neg\Box E \vee \neg\neg E),\\ &\supset \Box(\neg\Box E \vee \neg\neg E \vee G),\end{aligned}$$

and, as $\Box(\neg\Box E \vee E)$ is a substitutional instance of axiom C2, we establish as before, the provability of $\Box\chi(t)$, i.e. $\chi(t)^{\Box}$ in EpT.

The case, when $k > 0$, is considered likewise as in [1], that completes the proof of Lemma 2.

Further, as in [1] and [2] we state for the first-order predicate systems EpT and Ep4 that if $\Vdash A$, then $\vdash A$.

The constructions preparing the proof of Lemma 6 [1] for the propositional part of EpT and the proof of Lemma 10 [2] for the same part of Ep4, differ from each other only by the properties of the relations between tableaux. In [1] there was described the construction of a counter model and drafted the plan of proof of Lemma 6 [1]. And in [2] we have limited with a reference on [1].

These Lemmas will be proved for the predicate case parallelly below.

If a diagram for A is consistent, then such is also the main tableau t with the initial formula $\neg A$. All the auxiliary tableaux of t and at least one of the alternatives tableaux are consistent too, etc. Let's choose any alternative set $\mathcal{T}$ of consistent tableaux of the diagram of A. The main tableau of the diagram belongs to $\mathcal{T}$. Further, if $t \in \mathcal{T}$, then for every tableau t', which is auxiliary for $t, t' \in \mathcal{T}$. All these tableaux are consistent. Finally, if $t \in \mathcal{T}$ and t_1, t_2 are alternative ones for t, then $\mathcal{T}$ contains one of them, which is consistent too.

The set $\mathcal{T}$ is partially ordered by relation $\mathcal{R}$ between tableaux, which is reflexive in EpT and reflexive and transitive in Ep4 diagrams.

Now, by means of $\mathcal{T}$, let's define the model structure $S_0 = [H_0, W_0, R_0, D_0]$. An expression of the form $\overline{A}$ will be called nonessential marked, if formula A is fully modalized. We shall say that a tableau is unmarked, if among its initial formulas with upper lines may be only nonessential marked expressions.

Let θ be a function which transforms tableaux of $\mathcal{T}$ in the elements of H_0, i.e. θ is one-to-one mapping of $\mathcal{T}$ into H_0. Now, we can define W_0 as the subset of H_0, which contains all such elements $\theta(t)$ of H_0 that t is unmarked tableau (W_0 is non-empty, as, for instance, the main tableau is unmarked). The elements of H_0 are ordered by the relation R_0, which corresponds to the relation $\mathcal{R}$ between tableaux of $\mathcal{T}$, besides if $t_1, t_2 \in P\mathcal{T}, v_1 = \theta(t_1)$ and $v_2 = \theta(t_2)$, then $(v_1, v_2) \in R_0$ iff $(t_1, t_2) \in \mathcal{R}$.

It remains to define the domain function D_0 for S_0. If t is the main tableau of $\mathcal{T}$ and $v = \theta(T)$, then $D(_0(v) = X \cup X_t$ where X is the set of all free individual variables of the initial formula $\neg A$ of tableau t, and X_t is a subset of the set Y_t of individual variables, corresponding to t; and if A does not contain either free individual variables or quantifiers, then $D_0(v)$ is arbitrary non-empty subset X_t of the set Y_t. Let, further, $t_1, t_2 \in \mathcal{T}, v_1 = \theta(t_1), v_2 = \theta(t_2)$ and $(v_1, v_2) \in R_0$. Assume that we have already defined $D_0(v_1)$. Then $D_0(v_2) = D_0(v_1) \cup X_{t_2}$, where X_{t_2} is any subset of the set Y_{t_2}; if t_2 does not contain any formula with quantifiers, then X_{t_2} is arbitrary subset of the set Y_{t_2}. For every $v = \theta(t)$ and corresponding tableau t, in construction of which the rules of quantifiers $N\Pi$, $\overline{\Pi}$ are applied, X_t is the set of all new variables, which are introduced by the rules from Y_t. It is obvious that the condition (d) of domain function D_0 always will be held, as if $(v_1, v_2) \in R_0$, then $D_0(v_1) \subseteq D_0(v_1) \cup X_t, = D_0(v_2)$.

And, finally, EpT-, Ep4-model M_0 of the model structure S_0 (with reflexive, respectively, reflexive and transitive R_0) is a pair $[S_0, E_0]$. An evaluation function E_0 is connected with every tableau t from $\mathcal{T}$ by the following conditions. Let $v = \theta(t)$. When $n = 0$, $E_0(P^n, v) = T$, if t contains P^n, $E_0(P^n, v) = F$, if t contains $\neg P^n$; and non $!E_0(P^n, v)$, if t contains one of the expressions $\overline{P^n}$ or $\overline{\neg P^n}$, but in the first case $\neg P^n$ does not occur in t, and in the second case P^n does not occur in t (in particular, non $!E_0(P^n, v)$, if t contains both expressions $\overline{P^n}$ and $\overline{\neg P^n}$). And as to such a propositional letter P^0, for which t does not contain either P^0 or $\neg P^0$ and their absence in t is not marked by occurrence of the corresponding expression with the upper line, we suppose that $E_0(P^0, v)$ is any of truth values. This guarantees the definability of E_0 for any (propositional) formula and any element of W_0.

When $n > 0$, we define $E_0(P^n, v)$ as a partial function L_0 from $[D_0(v)]^n$ into T, F such that $L_0(x_1, \ldots, x_n) = T$, if $P^n(x_1, \ldots, x_n)$ occurs in t; $L_0(x_1, \ldots, x_n) = F$, if $\neg P^n(x_1, \ldots, x_n)$ occurs in t and L_0 is undefined for $x_1, \ldots, x_n$ if t contains one of the expressions $\overline{P^n(x_1, \ldots, x_n)}$ or $\overline{\neg P^n(x_1, \ldots, x_n)}$, however, in the first case $\neg P^n(x_1, \ldots, x_n)$, does not occur in t, and in the second case $P^n(x_1, \ldots, x_n)$ does not occur in t. In particular, non $!L_0(x_1, \ldots, x_n)$, if t contains both expressions $\overline{P^n(x_1, \ldots, x_n)}$ and $\overline{\neg P^n(x_1, \ldots, x_n)}$. Besides, we assume that $E_0(P^n, v)$ is arbitrary function from $[D_0(v)]^n$ into $\{T, F\}$ for all predicate letters P^n such that t contains either $P^n(x_1, \ldots, x_n)$ or $\neg P^n(x_1, \ldots, x_n)$ for no $x_1, \ldots, x_n$, and their absence is not marked in t by an occurrence of the corresponding expressions with the upper line. At the same time, this guarantees definability of E_0 for any formula and any element of W_0.

Now we can show that M_0 is a counter model for A, when the diagram for A is consistent.

As before, consider the union U_0 of all sets $D_0(v)$, where $v = \theta$ (T)4 and $t \in \mathcal{T}$. It is convenient to evaluate each formula of any tableau t for assignment by which to every of its free individual variables there corresponds the same variable from U_0, which (as we see) is identified with an object, i.e. when to each its free individual variable is assigned the variable itself as an object. ■

Preliminarily we state

Lemma 1.3 *Let M_0 be the model given above. Then for any formula B and any tableau t of an alternative set of consistent tableaux, when the variables from U_0 are identically assigned to all free individual variables of B, M_0 satisfies the following conditions:*

1. *If B occurs in t, then $E_0(B, v) = T$ for the given assignment,*
2. *If $\neg B$ occurs in t, then $E_0(B, v) = F$ for the given assignment,*
3. *If $\overline{\neg B}$ occurs in t, then non $!E_0(B, v)$ or $E_0(B, v) = T$ for the given assignment,*
4. *If $\overline{B}$ occurs in t, then non $!E_0(B, v)$ or $E_0(B, v) = F$ for the given assignment.*

PROOF. By simultaneous induction with respect to the number of the logical signs[2] of B.

When B is an atomic formula, then the conditions (1)–(4) hold immediately by the definition of E_0.

Suppose, hence, that formula B is not an atomic one. Then, taking into account that from every formula the signs $\&$ and $\exists$ can be eliminated, we shall limit our consideration by the cases, when B has one of the forms: $\neg C, C_1 \vee C_2, \forall x C(x)$ or $\Box C$.

Let B have a form $\neg C$. We shall show that the condition (1) holds. We use again the induction now with respect to the number of the logical signs of C, which in its turn may be an atomic formula or have one of the forms: $\neg D, D_1 \vee D_2, \forall x D(x)$ or $\Box D$. In case, when E is an atomic one, B has the form $\neg C$ and the condition (1) holds by the definition of E_0.

If C has a form $\neg D$, then B is the formula $\neg\neg D$ and, as it occurs in the consistent tableau t, the rule NN is applicable to it. According to this rule in t is also placed formula C, which contains one logical sign less than formula B. Therefore, by the inductive hypothesis $E_0(D, v) = T$ for the identical assignment the elements of U_0 to the free individual variables of D. But according to the evaluation condition for negation , $E_0(B, v) = E_0(\neg\neg D, v) = T$ for the same assignment.

Let C have a form $D_1 \vee D_2$. Then B is the formula $\neg(D_1 \vee D_2)$ and in the consistent tableau t the rule ND is applicable to it. According to this rule the formulas $\neg D_1$ and $\neg D_2$ are put in it. Each of these ones contains logical signs less than B. By the inductive hypothesis for the condition (2), $E_0(D_1, v) = E_0(D_2, v) = F$ for the given assignment. Whence, according to the evaluation conditions for disjunction and negation, it follows that $E_0(B, v) = E_0(\neg(D_1 \vee D_2), v) = T$ for the same assignment.

If C has a form $\forall x D(x)$, B is the formula $\neg\forall x D(x)$ and because of the consistency of t, the rule $N\Pi$ is applicable to the latter formula. According to this rule formula $\neg D(y)$ is put in t with earlier not occurring in any tableau free individual variable $y \in X_t \subseteq Y_t$. Besides, the number of the logical signs of $\neg D(y)$ is one less than that of B and, hence, we can apply the inductive hypothesis for the condition (2). In accordance with it, $E_0(D(y), v) = F$ for the identical assignment variables of U_0 to all different from y free individual variables of $D(y)$, when y from $X_t \subseteq D(v), v = \theta(t)$, is assigned to the free variable y of the formula $D(y)$. Whence, according to the evaluation condition for the universal quantifier, $E_0(\forall x D(x), v) = F$, and by the condition for negation, $E_0(B, v) = E_0(\neg\forall x D(x), v) = T$ for the given assignment.

Finally, if C has a form $\Box D$, then B is the formula $\neg\Box D$, and as t is consistent, the rule NK is applicable to the latter formula. According to this rule a new tableau t' starts, such that $(t, t' \in \mathcal{R})$ and the expression $\overline{D}$ is put in it. But D contains the logical signs one less than B and we can apply the inductive hypothesis for the condition (4). In accordance with it, non $!E_0 * D, w)$ or $D_0(D, w) = F$,

[2] I.e. the classical and modal signs.

where $w = \theta(t')$, for the identical assignment of elements of U_0 to all free variables of D. And then, according to the evaluation condition for the epistemic operator, $E_0(B, v) = E_0(\Box D, v) = F$ for the same assignment, as B and D contain the same free individual variable and $D_0(\theta(t)) \subseteq D_0(\theta(t'))$.

Thus, the condition (1) holds, when B has a form $\neg C$. It remains to consider the cases, when B has one of the forms $C_1 \vee C_2, \forall x C(x)$ or $\Box C$.

If B has a form $C_1 \vee C_2$, then, as t is consistent, the rule D is applicable to it and according to this rule we start out two alternatives for t tableaux, each of which entirely copies t and in addition contains one of the formulas C_1 or C_2. $\mathcal{T}$ contains one of these tableaux with ($I = 1$ or 2). It is consistent too. And as the number of the logical signs of C_i is less than that of B, by the inductive hypothesis, $E_0(C_i, v) = T$ for the identical assignment of elements of U_0 to the free variables of C_i and, according to the evaluation condition for disjunction $E_0(B, v) = E_0(C_1 \vee C_2, v) = T$ for the same assignment.

If B has a form $\forall x C(x)$, then, as t is consistent, the rule Π is applicable to the latter formula and for each individual variable y of every tableau t', such that $(t', t) \in \mathcal{R}$ and hence, for each free variable y of t in t is put formula $C(y)$ with the less than B number of the logical signs. Therefore, by the induction hypothesis, $E_0(C(y), v) = T$ for the identical assignment of elements of U_0 to all distinct from y free individual variables of $C(y)$, when any variable from $D_0(v), v = \theta(t)$, is assigned for the free variable of y of $C(y)$. But then according to the evaluation condition for universal quantifier, $E_0(B, v) = E_0(\forall x C(x), v) = T$ for the identical assignment variables of U_0 to all free variables of B.

Finally, if B has a form $\Box C$, then the rule K is applicable to it, as t is consistent, and in every tableau t', such that $(t, t') \in \mathcal{R}$, is placed the formula C with the less than B number of the logical signs. Hence, by the inductive hypothesis, for every w, such that $(v, w) \in R_0, v = \theta(t), w = \theta(t'), E_0(C, w) = T$ for the given assignment of the elements of U_0 to all free variables of C. But then, in accordance with the evaluation condition for the epistemic operator (B and C contain the same free variables), $E_0(B, v) = E_0(\Box C, v) = T$ for the same assignment.

So, the condition (1) of Lemma 1.3 is established.

The condition (2) is considered quite similarly as the condition (3) and we shall state here the latter, again by induction with respect to the number of the logical signs of formula B.

Let B have a form $\neg C$. Then, t contains an expression $\overline{\neg\neg C}$ and as t is consistent, to the latter formula the rule $\overline{NN}$ is applicable, according to which C is placed in t. By the inductive hypothesis of the condition (4), non $!E_0(C, v)$ or $E_0(C, v) = F$ for the identical assignment of the objects of U_0 to the free variables of C. Whence, according to the evaluation condition for negation, it follows that non $!E_0(B, v)$ or $E_0(B, v) = E_0(\neg C, v) = T$ for the same assignment.

If B has a form $C_1 \vee C_2$, then the consistent tableau t contains an expression $\overline{\neg(C_1 \vee C_2)}$. The rule $\overline{ND}$ is applicable to it and according to this rule we start out two alternative tableaux, each of which copies entirely t and in addition contains one of the expressions $\overline{\neg C_1}$ or $\overline{\neg C_2}$. Besides, one of these tableaux with the expressions $\overline{\neg C_i}$ $(i = 1$ or $2)$ is consistent and belongs to $\mathcal{T}$. But as the number of the logical signs of $\overline{\neg C_i}$ is less than that of B, by the inductive hypothesis, non $!E_0(C_i, v)$ or $E_0(C_i, v) = T$ for the identical assignment of elements of U_0 to the free variables of C_i. And then, according to the evaluation condition for disjunction, non $!E_0(B, v)$ or $E_0(B, v) = E_0(C_1 \vee C_2, v) = T$ for the same assignment.

Let us assume, now, that B has a form $\forall x C(x)$. Then, t contains the expression $\overline{\neg \forall x C(x)}$ and the rule $\overline{N\Pi}$ is applicable to the latter one. According to this rule for each free individual variable y of every tableau t', such that $(t', t) \in \mathcal{R}$, the expressions $\overline{\neg C(y)}$ are put in t. As the number of the logical signs of $C(y)$ is one less than that of B, by the inductive hypothesis, non $!E_0(C(y), v)$ or $E_0(C(y), v) = T$ for the identical assignment variables of U_0 to all distinct from y variables of $C(y)$, within any variable from $D_0(v), v = \theta(t)$, is assigned to the y. But then, according to the evaluation condition for the universal quantifier, in the case, when for the same assignment we have $D_0(C(y), v) = T$ for any assigned to y variable from $D_0(v), E_0(B, v) = E + 0(\forall x C(x), v) = T$, otherwise, non $!E_0(B, v)$.

Finally, let B have a form $\Box C$. Then, consistent tableau t contains the expression $\overline{\neg \Box C}$ and the rule $\overline{NK}$ is applicable to it. At first, according to this rule $\neg\neg\Box C$ is put in t, then the formula $\Box C$ is put in g in accordance with the rule NN, and after that in every tableau t', such that $(t, t') \in \mathcal{R}$, the formula C with one less than B number of the logical signs is placed in t according to the rule K. Therefore, to every of such tableau t' and to formula C we may use the inductive hypothesis for the condition (1), according to which $D_0(C, w) = T$ for every w, such that $(v, w) \in R_0$, for the identical assignment of the elements of U_0 to all free variables of $C, v = \theta(t), w = \theta(t'), D_0(v) \subseteq D_0(w)$. But by the evaluation condition for the epistemic operator, $E_0(b, v) = E_0(\Box C, v) = T$ for the same assignment, and therefore the condition (3) holds.

The condition (4) may be established quite similarly as the condition (1).

The lemma is proved.

As we see, it justifies semantically our rules for constructing tableaux.

Further, if a diagram for A is consistent, then its main tableau with the initial formula $\neg A$ is consistent too, whence, according to the condition (2) of Lemma 1.3, there is a counter model M_0 for A and, therefore, A is invalid. Then we argue, as in proving of completeness of the propositional parts of EpT and Ep4.

This completes the proof of the completeness theorem for EpT and Ep4. ∎

Connection with Wang's Calculus of Partial Predicates

In [2] it was shown that the propositional part of Hao Wang's calculus of partial predicates is expressible in the systems there investigated. This result can be extended on the predicate case. Though we limit our consideration by the system Ep4, as on its base an alternative system can be formulated, with permits iterations of Wang's implication.

The initial logical signs of Wang's calculus of partial predicates PP [7] are $\neg, \&, \vee, \forall, \exists$ and $\rightarrow$ (the sign of Wang's implication). The alphabet of PP contains also individual variables, predicate letters (may contain also functional signs, etc.). These signs may occur in Ep4 as well, however, in the considered cases no individual constant should coincide with the personal subscript of the epistemic operator). Every formula of PP is either '$\rightarrow$' free, or has a form $B \rightarrow C$ with '$\rightarrow$' free B and C.

Evaluation of '$\rightarrow$' free formulas is carried out by means of Łukasiewicz's three-valued tableaux, the universal and existential quantifiers are interpreted naturally as generalized conjunctions and disjunctions respectively, and finally, the formula $B - C$ is valid iff C takes the truth-value T, whenever B takes the value T for every evaluation.

Let correspond to each formula A of PP the formula $\tau(A)$ (translation of A) in the following way:

$\tau(A) = \Box A$, if A is '$\rightarrow$' free, and

$\tau(A) = \tau(B) \rightarrow \tau(C)$, if A has the form $B \rightarrow C$ with '$\rightarrow$' free in B and C.

It is easy to establish

Lemma 1.4 *If* $\models \tau(A)$ *in Ep4, then* $\models A$ *in* PP.

(For its proof it is sufficient to show that if A is refutable in PP, then $\tau(A)$ is refutable in Ep4).

Hence, we may establish that $\vdash A$ in PP iff $\vdash \tau(A)$ in Ep4 (cf. Theorem 5 [2]).

Actually, by immediate verifying we can easily be convinced that the translation of each axiom of PP is a theorem of Ep4, and the translation of each rule of inference of PP gives a derived rule of Ep4 (the propositional part of PP was adequately axiomatized by A. Rose in [6], and the postulates of the properly predicate part of PP are formulated in [7]). Therefore, if $\vdash A$ in PP, then $\vdash \tau(A)$ in Ep4. On the other hand, if $\vdash \tau(A)$ in Ep4, then because of consistence of the latter system, $\models \tau(A)$ in Ep4, whence, according to Lemma 1.4 $\models A$ in PP, and because of completeness of PP (see Theorem 1 [7]), $\vdash A$ in PP.

Possible applications

As we have already seen, the formula $A \rightarrow B$ with Wang's implication can be defined in Ep4 by $\Box A \supset \Box B$. Consequently, Ep4 enables us, factually, to consider the formulas which correspond to iterations of Wang's implication, including the iterations in antecedent. This is impossible for PP because, as it is noted in [7], the interpretation of the sign $\rightarrow$ there does not presuppose such cases. Can we evaluate in the such way, for example, the formula $(\neg P \rightarrow P) \rightarrow P$? In order that it be T, we require that for all evaluations, if $(\neg P \rightarrow P)$ takes the value T, then P must take the same value. But we have given no value to $\neg P \rightarrow P$, because it was not explained what does it mean for the latter formula to get the value T under each single evaluation. This is the reason of imposing restrictions on formulas of PP (see [7]).

In Ep4 we may define $A \rightarrow B$ as $(\Box A \supset \Box B)\&(\Box B \supset \Box A)$, but this will not be an equivalence, which guarantees the replacement of suitable translations of formulas of PP in Ep4. For this reason we can define $A \Leftrightarrow B$ in Ep4, as equivalence in this new sense, by $(\Box A \equiv \Box B)\&(\Box\neg A \equiv \Box\neg B)$. It is also obvious that Ep4 enables us to consider iterations of signs $\leftrightarrow$ and $\Leftrightarrow$. But in spite of such expressive possibilities, we cannot consider Ep4 as a relative system to $\dot{P}P$ because Ep4 contains all classical (nonmodalized) tautologies, but none of them is valid in PP.

Thus, if we wish to find out the alternative way of constructing an ideologically close to PP system, admitting iterations of Wang's implication, then either restricting the class of formulas of Ep4 by fully modalized ones, which will make the formulations of the evaluation condition for an epistemic operator somewhat complicated, or modifying model structures of Ep4, will be required. The second way seems more natural.

The alphabet and the language remains unchanged. In a model structure of the new epistemic first-order predicate system EIP we need not distinguish from H the non-empty set W any more. It is sufficient to consider non-empty set H with the same R and D, as in Ep4-model structure. Now, truth of a formula in a model and its validity in EIP are usually defined as being true for every element of H. It is obvious that no nonmodalized formula and consequently no such classical tautology will be valid in EIP.

A formula of EIP will be considered as a theorem, if it is valid in EIP. The problem of axiomatisability of EIP we have not investigated yet.

The signs $\rightarrow$, $\leftrightarrow$ and $\Leftrightarrow$ in EIP are introduced by the same definitions as in Ep4.

EIP has properties which may evoke some interest in regard to one of its extensions.

Notice that in the formulations of systems in [1, 2] and [3] partial descriptions of possible worlds were used as the semantical basis which enables us to avoid the paradox of logical omniscience. Besides, these systems were built with such an intention that all classical tautologies remained theorems. Unlike them EIP does not contain any nonmodalized classical tautology and the formulas A and $-A$ are not necessarily contradictory here. They may be equally undefined. These peculiarities may prove to be decisive for avoiding more undesirable paradoxes than that of logical omniscience. As it is known, D. Bochwar used similar peculiarities of his three-valued logic to eliminate logical and semantically antinomies constructing extended predicate calculus free from type theoretical restrictions (see e.g. [4]).

In an analogous manner and with the same intention we can extend EIP introducing into consideration predicates having as arguments propositions and predicates without any limitations of types (argument values of each predicate belong to some domains accordingly extended). Let EEP be such an extension of EIP. In the meantime we can only suppose that it is consistent. The empirical grounds for this supposition is the fact that in EEP we cannot reproduce the antinomies for the reasons analogous to those considered in [4]. Besides, the comprehension axiom formulated in terms of $\Leftrightarrow$ does not lead us to Russell's antinomy. At any rate, EEP should not be trivially contradictory. But if it is really consistent, then we could test the logistic thesis once more on the new basis of the extended epistemic predicate system without type restrictions.

Georgian Academy of Sciences, Tbilisi

References

[1] M. H. Bezhanishvili. Some epistemic propositional systems. *Metody logicheskikh issledovanij*, pp. 27–38. Tbilisi. In Russian, 1987.

[2] M. H. Bezhanishvili. Epistemic logic and semantics of partial description of 'possible worlds'. *Issledovanie neklassicheskikh logik*. Nauka, Moskow. In Russian, 1989.

[3] M. H. Bezhanishvili. Dxatic propositional systems with single modal operator. *Intensional'nye logiki i logicheskaja struktura teorij*. Tbilisi. In Russian, 1988.

[4] D. A. Bochvar. On a three-valued calculus and its application to analysis of paradoxes of extended classical functional calculus. *Matematicheskij sbornik*, **4**, (46), 1938. In Russian.

[5] J. Hintikka. *Knowlege and Belief. Introduction to the Logic of Two Notions*, Cornell University Press, Ithaca, 1962.

[6] A. Rose. A formulation of th epropositional calculus corresponding to Wang's calculus of partial predicates. *Zeitschrift für mathematische Logik und Grundlagen der Mathematik*, **9**, 177–198, 1963.

[7] Wang Hao. The calculus of partial predicates and its extension to set theory I. *Zeitschrift für mathematische Logik und Grundlagen der Mathematik*, **7**, 283–288, 1961.

13

Normative Reasonings and Default Assumptions

Irina Gerasimova

The very availability of patterns and standards of mental activities rooted in social experience allows us to reduce detailed analysis in every particular case of individual thinking and use the implicit and tacit mechanisms of cognizing. This specific feature of human thinking lies in the basis of so called reasonings by default. In everyday life a man often makes inferences according to certain rules if there are no specific indications or exceptions to these rules.

From the logical viewpoint, reasonings by default can be essentially considered as forms of nonmonotonic inference in which new axioms can refute former theorems.

Here I would like to pay attention to the field of practical activities where use of reasoning by default, seems quite natural. We often use such reasonings when we deal with norms, instructions, rules regulating behaviour and in making a decision. Since it is impossible to regiment actions strictly in all possible situations, the codes of norms would, normally, explicitly register only some essential points, and they would implicitly allow the actions which are consistent with the code in question. An absence, from the code, of prohibition to do something means a permission to do this.

In the present article deontic as well as epistemic norms such as 'It is forbidden not to know' will be considered. The new intensional operators for description of inferences by default will be introduced with a formal semantics of allocative system type.

In many cases the default reasonings can be expressed by an UNLESS- connective. For example:

(a) A bird can fly unless it is a penguin.

(b) A sick man can take a medicine unless this medicine is contra- indicated to him.

Let 'pUq' is 'p unless q'. If we interpret [2] UNLESS- connective as the exclusive disjunction then the following holds:

P. I. Bystrov and V. N. Sadovsky (eds.), Philosophical Logic and Logical Philosophy, 203–212.

1. $((pUq)\&\neg q) \supset p$

2. $((pUq)\&p) \supset \neg q$

Under an extentional approach the lack of information about something is unequal to the ontological assertion to the effect that it is absent in reality. In classical logic there is only one type of negation which covers cases of a lack of information.

Let's consider as an example the following situation. A sick man comes to a doctor and complains of a chill. The doctor examines him, asks questions, studies his history of illness and, finally, recommends him to take aspirin pills because he finds nothing which would be contra-indicated to the organism of his patient. But using this medicine leads the patient to a grave condition. Perhaps, our fears of aspirin are vain but there is a great number of other medicines which are not so safe as they seem. Moreover, the experts' opinion about his actions are constantly changeable. In any case, it would be better to be careful about substituting the assertion that p does not take place for the lack of information about state of affairs p.

Finnish logician Antti Hautamäki suggested an intensional interpretation of UNLESS-connective in terms of epistemic logic [1]:

3. $[(pUq)\&(\neg pU\neg q)\&\neg Kq] \supset \neg K\neg p.$

If we consider only K-operator and UNLESS-connective, then it may be enough to use UNLESS in the scope of K:

4. $(K(pUq)\&\neg Kq) \supset \neg K\neg q.$

This treatment of UNLESS stimulates the further discussion. Let Dp denote 'by default p'. It is necessary to ask the following questions:

(a) Can D be defined in terms of K?

$$Dp \leftrightarrow \neg K\neg p? \qquad \text{or} \qquad \neg K\neg p \rightarrow Dp?$$

To answer this question we need to consider the following problems:

(b) how can one's knowing something be represented?

(c) How can one's not knowing something be represented?

Let's start from the latter question. I suppose knowing and not knowing to be complicated modalities. Accordingly, an agent's knowledge can be interpreted, for instance, in the following way: 'an agent knows that p' means 'he believes that p, can sufficiently justify that p and p is true'. The latter condition means that p is true in some fixed system which can serve as representation of an observer (investigator). In symbols:

5. $Kp =_{\text{df}} Bp\&Jp\&p.$

Now consider the following equivalences based on the definition (5):

6. $\neg Kp \leftrightarrow \neg(Bp\&Ip\&p)$ or $\neg Kp \leftrightarrow (\neg Bp \vee \neg Jp \vee \neg p)$.

The last says that 'agent doesn't know that p' means that either 'he doesn't believe that p' ($\neg Bp$ holds) or 'he is mistaken in believing that p' ($Bp\&\neg p$ holds) or 'he is mistaken in believing that he knows that p' ($Bp\&Jp\&\neg P$ holds). Furthermore, the agent's not believing that p can indicate two possibilities: either he believes that $\neg p$ or he believes neither p nor $\neg p$ in the case of a value gap. In the latter case we deal with the agent's lack of information about p. Thus under this interpretation of knowledge, the lack of information (knowledge) is a particular case of unknowingness.

Usually we never say 'I don't know that it is raining' but 'I don't know whether it is raining or not'. Under autoepistemic treatment the whole proposition has interrogative or problematical character. There is another situation in the case of the third person, namely an absence of information about p is relative. It is implicated that agent doesn't know something known by observer. Then the whole sentence has affirmative form: 'agent doesn't know that p'. Note that formula $\neg Kp \supset K\neg Kp$ is appropriate for autoepistemic logic and is not appropriate for epistemic logic of the third person because there is a case when an agent doesn't think about information an observer has.

This consideration shows that it is not easy to transfer Dp to $\neg K\neg p$ because of interrogative meaning of $\neg p$ in $\neg K\neg p$. But such transformation is valid under certain conditions.

If we would like to deal with a rational agent we must take into account such characteristics of knowledge as justification and truthfulness. The agent has to justify his knowledge not only for himself but also for observer. Only under such conditions can the observer justify the fact of an agent's known that p. Truthfulness in that case means *truthfulness relative to a fixed system.* For instance, the fixed system can represent a common point of view. In other words we can have a look at knowledge through social dimension. There is a great number of norms, directions, regulations, facts which exist independently from the individuals and which individuals can use in their reasoning. In this case epistemic logic is proved to be like conditional deontic logic. The last describes how norms can be used under different conditions. Analogously, epistemic logic can describe how epistemic norms are used in cases of incomplete information: when one parameter is known and another is not.

Thus the main idea is to introduce the normative epistemic operators and the special operators of the lack of information for treating a particular kind of unless-norms such as deontic ones.

Constructing language and its semantics we shall distinguish assertions of external and internal levels. The propositional language L including Boolean connectives $\&, \vee, \supset, \neg$ and $\backslash$ (UNLESS) is used for description of internal level. Propositions of this type are used for description of states of affairs and norms codes. Let L_e be a language which includes propositions of external level (propositional and modal assertions). The external level represents observer's reasonings about using norms under conditions when some circumstances are known and another are not.

Let *Var* be a fixed set of propositional variables $p, q, r, \ldots$, which represent propositions to describe states of affairs or processes that can or cannot take place (continue) at a certain time or under certain circumstances. The set of well-formed formulas of language L is the smallest set *Form*(L) such that:

1. a propositional variable $p \in Form(L)$;
2. if $A \in Form(L)$ and $B \in Form(L)$, then $(A\&B) \in Form(L)$, $(A \vee B) \in Form(L)$, $(A \supset B), \neg A \in Form(L)$.
   All formulas described in (1) and (2) will be called propositional formulas.
3. If A and B are propositional formulas then $(A \backslash B) \in Form(L)$.

Negation $\rceil$, conjunction $\cap$, disjunction $\cup$ and implication $\rightarrow$ are logical connectives of external level as well as the binary intensional operator UNLESS ($\backslash$). The vocabulary of external language L_e includes the following operators too.

1. Deontic operators
   P ('It's permitted that'), *notP* ('It's not permitted that'), F ('It's forbidden that'), *notF* (It's not forbidden that'), IP (It's implicitly permitted that'), DP ('It's permitted by default that' or 'It's permitted in the absence of contra-indications that'), $\Diamond P$ ('It's possibly permitted that').
2. Operator of Epistemic norm:
   $F \sim$ ('It's forbidden not to know that');
3. Epistemic operator of the lack of information $\sim$.

A set of well formed formulas of external language is the smallest set *Form*(L_e) such that:

1. If A is a propositional formula of the set *Form*(L) then $A \in Form(L_e)$.
2. If A is a propositional formula of the set *Form*(L) then PA, *notP*A, FA, *notF*A $\in Form(L_e)$. These formulas are called deontic formulas of the first level.
3. If A is a propositional formula of the set *Form*(L) then IPA, DPA, $\Diamond PA \in Form(L_e)$. The last formulas are deontic formulas of the second level.
4. If A and B are propositional formulas of *Form*(L) then $PA\backslash B$, $FA\backslash B \in Form(L_e)$, which are called derivative deontic formulas of the first level.
5. If A is a propositional formula of *Form*(L) then $F \sim A \in Form(L_e)$. $F \sim A$ is an epistemic norm.
6. If δ is a propositional deontic formula of the first level, a derivative deontic formula or an epistemic norm then an epistemic formula $\sim \delta \in Form(L_e)$.

7. If φ and $\psi \in Form(L_e)$ then $\varphi \cap \psi, \varphi \cup \psi, \varphi \rightarrow \psi, \rceil\varphi \in Form(L_e)$.

In all modal formulas operators are understood as identificators or whole symbols even if they are of considerable complexity. Each operator is a certain valuation of a state or a process. For example, $F \sim A$ means 'it's forbidden not to know whether A is true or not' (or 'it's forbidden not to know a truth value of A'). The whole symbol $F \sim$ denotes such evaluation as 'it's forbidden not to know'. In other words, $F \sim$ is not definable in terms of the deontic operator F and the epistemic operator $\sim$.

A syntactic interpretation of all modalities in an allocative system will be given below. It means that modalities will be understood as specific predicates on propositions. For example, PA means that A is an element of a set of propositions which is a domain of P (in correspondent point of reference).

Note that from the definition of well formed formula of L_e follows that iterated modalities are not considered. For example, PPA, $F \sim PFA$ are not formulas of L_e.

Modality $\sim$ indicates a lack of information. This fact is fixed in the language of external level (L_e). Let A be a propositional formula of L_e. $\sim A$ means that there is no information about truth value of A. $\sim PA$ means that there is no information about deontic status of permission of A. $\sim F \sim A$ expresses the fact that there is no information whether prohibition of not knowing that A is an epistemic norm or not.

If a proposition A describes a situation then the modality 'it's forbidden not to know' will indicate a system (a normative code) in which this description is used. We shall not use the explicit negation of proposition 'it's forbidden not to know' (formally, *notF* $\sim A$) because in natural language the proposition 'it's forbidden not to know' is usually understood as lack of explicit indication in a correspondent normative code. In other words the epistemic proposition $\sim F \sim A$ is used more often than ontological form *notF* $\sim A$. It will be shown below that a lack of information about the epistemic normative status of proposition will have another meaning than the lack of information about its truth or deontic status.

Let's call norms P and F positive and *notP* and *notF* negative accordingly. The operator of not permitting (*notP*) is not directly connected with the operator of permitting (P), the same concerns to the pair of forbidding (F) and not forbidding (*notF*). This means that the above notions are thought to be relatively independent of each other. So not permitting is not treated as a simple negation of permitting. Both domains of permitting and not permitting constitute a system of weak norms, and domains of forbidding and not forbidding constitute a system of strong norms. The distinction between strong and weak normative systems is that the strong normative system is considered as 'the highest instance'. Carring out of a strong norm is preferred to carring out of a weak one. Sanctions for breaching a strong norm can be stricter than in the case of a weak one.

Formulas $PA \backslash B$ and $FA \backslash B$ represent norms with reserves. They look very much like conditional norms PA/B ('A is permitted under condition B') and

FA/B ('A is forbidden under condition B'). The distinct feature of norms with reserve is the following. Permitting (or forbidding) in norms with reserve indicates carring out a norm in most cases (or at least indicates a desirability of such carring out) and transforms into the contrary requirement in circumstances mentioned in the reserve. Under my interpretation the connective UNLESS will get meaning of intensional exclusive-or operator.

Let's give an interpretation of multimodal language L_e. For that purpose the idea of interpretation of modal languages in an allocative system will be explored. By the latter I understand a complex system (a state machine) consisting of subsystems connected in a certain way. Each of such subsystems is in a certain state in a fixed point of reference. A global state is an exhaustive complete description of states of all its subsystems.

The allocative system (let's call it $\mathcal{M}$) consists of five data blocks and the governor block (processor). Data blocks contain information about state of affairs and processes. On the basis of these data the governor block values permitting or forbidding of certain states or actions and makes conclusions. In particular it can transform data of blocks by using a new information. The internal language is used for description of data blocks and the external language L_e describes activities of the governor block $\mathbb{S}$.

All information is allocated in datablocks in the following way. The weak norms take place in block $\mathbb{P}$, the strong ones are in block $\mathbb{F}$, block $\mathbb{F}\ \sim$ contains epistemic norms, block $\mathbb{T}$ contains data facts. The additional block $\mathbb{H}$ is connected with blocks $\mathbb{P}$ and $\mathbb{F}$ and it contains conditions of carrying out the derivative deontic norms. Each block can include one positive and one negative section which will be denoted by $+$ and $-$ accordingly. In $\mathbb{P}^+$ there are norms of permitting, in $\mathbb{P}^-$ norms of not permitting; norms of forbidding are in $\mathbb{F}^+$, norms of not forbidding are in $\mathbb{F}^-$. Block $\mathbb{F}^\sim$ contains only the positive section $\mathbb{F}^\sim$ of epistemic norms of forbidding. $\mathbb{T}^+$ represents all what is actually true, $\mathbb{T}^-$ represents all what is actually false. $\mathbb{H}^+$ represents all what's conditionally true and $\mathbb{H}^-$ represents all what's conditionally false.

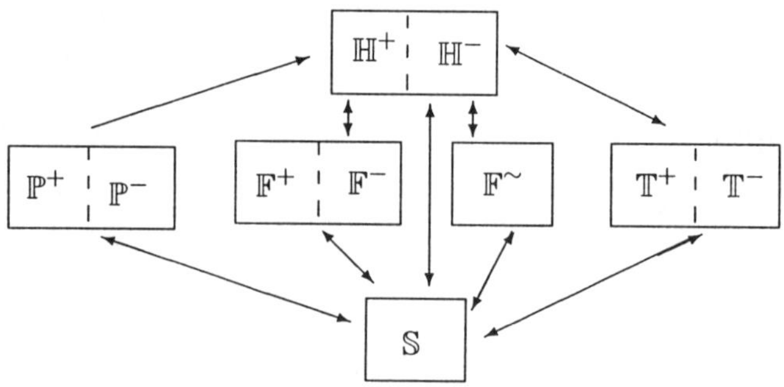

The allocative system $\mathcal{M} = \langle \mathbb{W}, \mathbb{P}, \mathbb{F}, \mathbb{F}^\sim, \mathbb{H}, \mathbb{T}, \mathbb{S}\rangle$ is an interpretation of multimodal language L_e if $\mathbb{W}$ is a fixed set of global states and $\mathbb{P}, \mathbb{F}, \mathbb{F}^\sim, \mathbb{H}, \mathbb{T}$ is the partial functions on $\mathbb{W}$ such that for any datablock i, for any global state $w \in \mathbb{W}i(w)$ is a set of formulas of propositional language L. In other words $i : \mathbb{W} \rightarrow 2^{Form(L)}$. The governor block $\mathbb{S}$ is a function from $\mathbb{W}$ into a set of formulas of the external language L_e. Formally: $\mathbb{W} \rightarrow 2^{Form(L_e)}$.

Let i be one of data blocks $\mathbb{P}, \mathbb{F}, \mathbb{F}^{\sim}, \mathbb{H}, \mathbb{T}$. A stroke $'$ is either $+$ or $-$. For arbitrary i, for any propositional variable p either $p \in i(w)$ or $p \notin i(w)$. If $p \in i(w)$ then either $p \in i^+_w$ or $p \notin i^-_w$.

The following conditions hold for complex propositional formulas:

1. if $(A\&B) \in i'_w$ then $A \in i'_w$ and $B \in i'_w$;
2. if $(A \vee B) \in i'_w$ then either $A \in i'_w$ or $B \in i'_w$;
3. if $(A \supset B) \in i'_w$ then either $\neg A \in i'_w$ or $B \in i'_w$;
4. if $\neg(A\&B) \in i'_w$ then $\neg A \in i'_w$ or $\neg B \in i'_w$;
5. if $\neg(A \vee B) \in i'_w$ then $\neg A \in i'_w$ and $\neg B \in i'_w$;
6. if $\neg(A \supset B) \in i'_w$ then $A \in i'_w$ and $\neg B \in i'_w$.

Intersection of i^+_w and i^-_w is empty: $i^+_w \cap i^-_w = \varnothing$. There is no formula A such that $A \in i^+_w$ and $A \in i^-_w$ at the same time. It's easy to notice that $\langle i^+_w, i^-_w \rangle$ is a generalised Hintikka's pair [3] for propositional formulas.

When A and B are propositional formulas of $Form(L)$, the connective $\backslash$ is defined by the following conditions:

1. If $A\backslash B \in \mathbb{P}^+_w$ then either $A \in \mathbb{P}^+_w$ and $B \in \mathbb{H}^-_w$ or $A \in \mathbb{P}^-_w$ and $B \in \mathbb{H}^+_w$.
2. If $A\backslash B \in \mathbb{F}^+_w$ then either $A \in \mathbb{F}^+_w$ and $B \in \mathbb{H}^-_w$ or $A \in \mathbb{F}^-_w$ and $B \in \mathbb{H}^+_w$.

Rules of block-normalization:

(A) if $A \in \mathbb{F}(w)$ then $A \notin \mathbb{P}(w)$;

(B) if $A \in \mathbb{F}^-_w$ then $A \cup \mathbb{P}^+_w$ does not lead to contradiction in $\mathbb{P}^+_w$.

Now we shall define the possible states of the governor block $\mathbb{S}$. Let w be an arbitrary global state of $\mathbb{W}$.

A Evaluations of state of affairs on the ground of indications of one of data-blocks in w:

1. If $A \in \mathbb{P}^+_w$ then $PA \in \mathbb{S}(w)$;
   if $A \in \mathbb{P}^-_w$ then $notPA \in \mathbb{S}(w)$;
   if $A \notin \mathbb{P}^+_w$ and $A \notin \mathbb{P}^-_w$ then $\sim PA \in \mathbb{S}(w)$;
   if $A\backslash B \in \mathbb{P}^+_w$ then $PA\backslash B \in \mathbb{S}(w)$.

B If $A \in \mathbb{F}_w^+$ then $FA \in \mathbb{S}(w)$;
if $A \in \mathbb{F}_w^-$ then *not*$FA \in \mathbb{S}(w)$'
if $A \notin \mathbb{F}_w^+$ and $A \notin \mathbb{F}_w^-$ then $\sim FA \in \mathbb{S}(w)$;
if $A\backslash B \in \mathbb{F}_w^+$ then $FA\backslash B \in \mathbb{S}(w)$.

C If $A \in \mathbb{F}^{\sim}(w)$ then $F \sim A \in \mathbb{S}(w)$;
if $A \notin \mathbb{F}^{\sim}(w)$ then $\sim F \sim A \in \mathbb{S}(w)$.

D If $A \in \mathbb{T}_w^+$ then $A \in \mathbb{S}(w)$;
if $A \in \mathbb{T}_w^-$ then $\neg A \in \mathbb{S}(w)$;
if $A \notin \mathbb{T}_w^+$ and $A \notin \mathbb{T}_w^-$ then $\sim A \in \mathbb{S}(w)$.

II Evaluation of state of affairs according to the indications of different data-blocks.

A If $PA\backslash B \in \mathbb{S}(w)$ and $\neg B \in \mathbb{S}(w)$ then $PA \in \mathbb{S}(w)$;
if $PA\backslash B \in \mathbb{S}(w)$ and $B \in \mathbb{S}(w)$ then *not*$PA \in \mathbb{S}(w)$;
if $PA\backslash B \in \mathbb{S}(w)$ and $\sim B \in \mathbb{S}(w)$ then if $F \sim B \in \mathbb{S}(w)$ then *not*$PA \in \mathbb{S}(w)$;
if $PA\backslash B \in \mathbb{S}(w)$ and $\sim B \in \mathbb{S}(w)$ then if $\sim F \sim B \in \mathbb{S}(w)$ then $DPA \in \mathbb{S}(w)$.

B If $FA\backslash B \in \mathbb{S}(w)$ and $\neg B \in \mathbb{S}(w)$ then $FA \in \mathbb{S}(w)$;
if $FA\backslash B \in \mathbb{S}(w)$ and $B \in \mathbb{S}(w)$ then *not*$FA \in \mathbb{S}(w)$;
if $FA\backslash B \in \mathbb{S}(w)$ and $\sim B \in \mathbb{S}(w)$ then $FA \in \mathbb{S}(w)$.

C The rule of rational obligation: if $\sim PA \in \mathbb{S}(w)$ and $\sim FA \in \mathbb{S}(w)$ and if B and $\neg B \in \{\neg A \cup \mathbb{P}(w)\}$ for some B then $\Diamond PA \in \mathbb{S}(w)$.

III Additional evaluation.
If *not*$FA \in \mathbb{S}(w)$ then $IPA \in \mathbb{S}(w)$.

Let's define the notion of truth in a point $w \in \mathbb{W}$ of the model $\mathcal{M}$. Let δ be any formula defined in the cases 1–6 of definition of *Form*(L_e), φ and ψ arbitrary formulas of language L_e.

A formula φ is true in the model $\mathcal{M}$ in a point $w \in \mathbb{W}(\mathcal{M}, w \models \varphi)$:

$\mathcal{M}, w \models \delta$ iff $\delta \in \mathbb{S}(w)$;
$\mathcal{M}, w \models \varphi \cap \psi$ iff $\mathcal{M}, w \models \varphi$ and $\mathcal{M}, w \models \psi$;
$\mathcal{M}, w \models \varphi \cup \psi$ iff $\mathcal{M}, w, \models \varphi$ or $\mathcal{M}, w \models \psi$;
$\mathcal{M}, w \models \varphi \rightarrow \psi$ iff if $\mathcal{M}, w \models \varphi$ then $\mathcal{M}, w \models \psi$;
$\mathcal{M}, w \models \rceil\varphi$ iff $\mathcal{M}, w \not\models \varphi$.

A formula φ is true in the model $\mathcal{M}$ iff for all $w \in \mathbb{W}$ $\mathcal{M}, w \models \varphi$. A formula φ is valid in a class of models of a given type iff for all values of functions $\mathbb{P}, \mathbb{F}, \mathbb{F}^{\sim}, \mathbb{T}, \mathbb{H}, \mathbb{S}$ for all $w \in \mathbb{W}$ $\mathcal{M}, w \models \varphi$.

The above described semantics of multimodal language enables us to show different meanings of what is permitted. Conditions of the modal operator P describe the explicit permission, i.e. just those norms that are fixed in the code. These conditions can be expressed by formulas like $P(A\&B) \rightarrow (PA \cap PB)$; $((\rceil PA \cap \rceil notPA) \rightarrow \sim (PA)$. It is easy to check that they are true in the model.

Other types of permission follow from the explicit not forbidding of some action or state. It is represented by the operator of implicit permission IP. Note that, in accordance with the rule of normalisation, if A is implicitly permitted then A does not lead to contradiction in the subsystem of what is explicitly permitted.

Not forbidding $\sim FA$ means a lack of explicit forbidding fixed in the code of norms. In this case a possible permission is introduced as lack of forbidding. In accordance with definition of the operator $\Diamond P$ permission is possible if it doesn't bring contradiction into the code of norms.

The peculiar type of permission is connected with norms with reserves. Conditions of connective UNLESS can be expressed by following schemes:

S1 $((PA \backslash B) \cap B) \rightarrow notPA$;

S2 $((PA \backslash B) \cap \neg B) \rightarrow PA$;

S3 $((PA \backslash B) \cap \sim B) \rightarrow (F \sim B \rightarrow notPA)$;

S4 $((PA \backslash B) \cap \sim B) \rightarrow (\sim F \sim B \rightarrow DPA)$.

S1 and S2 are classical conditions for exclusive disjunction ($\cup$), S3 and S4 describe situations of lack of information about parameter B, i.e. about reserve. Suppose that in some code there is permission to do A except cases when condition B appears. It may be a case where a lack of information about condition B. There is testing condition in interpretation: it is necessary to apply to 'the highest instance', i.e. to normative code, which is preferred to the other. If that code forbids not to know the reserve B then A is not permitted. If it is forbidden not to know B in the sense of lack of forbidding then A is permitted by default. Note that category 'it's forbidden not to know' marks out just the most essential conditions.

Let's remember the example with medicine. Usually it is permitted to use aspirin during a chill unless there are contra-indications to it. Suppose that a rigorous head of a hospital wrote instruction which forbids not to know the most important contra-indications. Then it is not permitted to prescribe medicine until the additional examinations of certain parameters will be done. The parameters belonged to medicines which are strictly used only according to the doctor's prescriptions fall just under the category 'it's forbidden not to know'.

The following formulas express connective UNLESS for code of forbidding:

S5 $((FA \backslash B) \cap B) \rightarrow notFA$;

S6 $((FA \backslash B) \cap \neg B) \rightarrow FA$;

S7 $((FA \backslash B) \cap \sim B) \rightarrow FA$.

The code of forbidding norms represents 'the highest instance' and therefore the lack of information about the reserve ($\sim B$) is equated with the explicit negation ($\neg B$). For example, assume that it is forbidden to park a car near a very important building unless you have special permission. It may happen that a driver left this document at home. The demonstration of the document is obligatory and if the driver can't do it he will be forbidden to park his car near this important building.

Russian Academy of Sciences, Moscow

References

[1] A. Hautamäki. *Epistemic Logic and Default Reasoning in Knowledge-Based System*, Helsinki, Manuscript, 1987.

[2] R. S. Michalski and P. H. Winston. Variable precision logic. *Artificial Intelligence*, **29**, 121– 146, 1986.

[3] E. D. Smirnova. *Logicheskaya Semantika i Philosophskie Osnovaniya Logiki*, Moskwa, 1986, s. 93–101.

14

A Logic Without Fixed Points

A. S. Karpenko

To approach this kind of logic, let us begin with what might seem a rather far-fetched problem, namely: the interrelations of Łukasiewicz's matrix many valued logic [9] and Gödel's matrix many-valued logic [5] which happens to be a generalization of Heyting's matrix three- valued logic (see [15, p. 45]).

Consider the logical matrices for both logics:

$$\mathcal{M}_n^L = \langle M_n; \{1\}, \sim, \rightarrow\rangle \text{ and } \mathcal{M}_n^g = \langle M_n; \{1\}, \neg \Rightarrow, \vee, \wedge\rangle,$$

where $M_n = \{0, 1/n-1, \ldots, n-2/n-1, 1\}(n \in N, n \geq 2)$, $\{1\}$ is the set of designated elements of $\mathcal{M}_n^L(\mathcal{M}_n^g)$,

$$\begin{aligned}
&\sim x = 1 - x,\\
&x \rightarrow y = \min(1, 1 - x + y),\\
&x \Rightarrow y = \begin{cases} 1, & \text{if } x \leq y \\ y, & \text{if } x > y \end{cases}\\
&\neg x = x \Rightarrow 0,\\
&x \vee y = \max(x, y) = (x \rightarrow y) \rightarrow y,\\
&x \wedge y = \min(x, y) = \sim(\sim x \vee \sim y).
\end{aligned}$$

Moisil [11] showed in 1963 that three-valued Łukasiewicz algebras are Heyting algebras, i.e. the operations of $\mathcal{M}_3^g$ are defined in $\mathcal{M}_3^l$, but not vice versa.

For the case $w < n < \omega$ we have the following:

$$J_i(x) = \begin{cases} 1, & \text{if } x = i \\ o, & \text{if } x \neq i \end{cases} \qquad (0 \leq i \leq 1)$$

Note that $\neg x = J_0(x)$. Rosser and Turquette [16, pp. 18–22] showed that the operators $J_i(x)$ can be defined in terms of $\sim x$ and $x \rightarrow$. And Cignoli [3, p. 10] showed that

$$x \Rightarrow y = J_1(x \rightarrow y) \vee y,$$

where $J_1(x) = J_0(\sim x)$.

P. I. Bystrov and V. N. Sadovsky (eds.), Philosophical Logic and Logical Philosophy, 213–219.

The last case to be considered is when $n = \omega$, for example, M_{χ_0} is the set of all rational numbers from $[0, 1]$. From McNaughton's criterion [10] for expressibility of functions in $\mathcal{M}^L_{\chi_0}$ it follows that operation $x \Rightarrow y$ is unexpressible in $\mathcal{M}^L_{\chi_0}$.

Thus, the problem consists in constructing such a logical matrix (a model) in which operation $x \Rightarrow y$ is definable. Let it be noted at once that such a model will be essentially non-standard.

Consider the matrix

$$\mathcal{M}^*_{\chi_0} = \langle [0,1]^*; \{1\}, \sim, \rightarrowtail \rangle,$$

where the set $[0, 1]^*$ is the set $[0, 1]$ from which are eliminated all the numbers with even denominator and

$$x \rightarrowtail y = \begin{cases} x, \text{ if } 0 < x = y < 1 \\ x \to y, \text{ otherwise.} \end{cases}$$

Thus, the matrix $\mathcal{M}^*_{\chi_0}$ contains no fixed point relative to operation $\sim x$, that is it contains no such elements that $\sim x = x$.

Let $\mathcal{L}_1$ denote the set of matrix functions from $\mathcal{M}^*_{\chi_0} = \langle [0,1]^*; \{2\}, \sim, \rightarrowtail \rangle$ and $\mathcal{L}_2$ denote the set of matrix functions from $\mathcal{M}^*_{\chi_0} = \langle [0,1]^*; \{1\}, \sim, \neg, \to \rangle$. Then we have

Lemma 0.1

$$\mathcal{L}_2 \subseteq \mathcal{L}_1.$$

PROOF. To prove Lemma 0.1 we have to express operations $x \to y$ and $\neg x$ by means of superposition of $\sim x$ and $x \rightarrowtail y$. It can be done as follows:

$$\begin{aligned} A &= \sim((y \rightarrowtail x) \rightarrowtail \sim (y \rightarrowtail x)) \rightarrowtail (x \rightarrowtail y) = x \to y, \\ B &= \sim((x \rightarrowtail) \to x) = \neg x. \end{aligned}$$

Proof of (A):

1. $x < y$. Then $x \rightarrowtail y - 1$ and hence $A = 1 = x \to y$.

2. $x = y$.

   (a) $x = 0$ or $x = 1$. Then $x \rightarrowtail y = 1$, and hence $A = 1 = x \to y$.

   (b) $0 < x < 1$. Then $A = \sim (x \rightarrowtail \sim x) \rightarrowtail x$. We have two subcases

      i. $x < 1/2$. Then $A = \sim (x \rightarrowtail \sim x) \rightarrowtail x = 0 \rightarrowtail x = 1 = x \to y$.

      ii. $x > 1/2$. Then $A = \sim (x \rightarrowtail \sim x) \rightarrowtail x = (1-(1-x+1-x) \rightarrowtail x = (2x-1) \rightarrowtail x$. Since $(2x-1) < x$ so $(2x-1) \to x = 1$. Hence $A = 1 = x \to y$.

3. $x > y$. Then $y \rightarrowtail x = 1$. Hence $A = \sim (1 \rightarrowtail 0) \rightarrowtail (x \rightarrowtail y) = x \rightarrowtail y = 1 - x + y = x \to y$.

Proof of (B):

1. $0 < x < 1$. Then $x \rightarrowtail x = x$, but $x \to x = 1$. Hence $B = \sim 1 = 0 = \neg x$.
2. $x = 0$. Then $B = \sim (1 \to 0) = 1 = \neg x$.
3. $x = 1$. Then $B = \sim (1 \to 1) = 0 = \neg x$.

This implies that operation $x \Rightarrow y$ also is expressible in matrix $\mathcal{M}^*_{\chi_0}$:

$$x \Rightarrow y = J_1(x \to y) \vee y.$$

■

Lemma 0.2

$$\mathcal{L}_1 \subseteq \mathcal{L}_2.$$

PROOF. It can be done as follows:

$$C = ((J_1(x \equiv y) \to x) \wedge (x \to y)) \vee \neg x = x \rightarrowtail y,$$

where $x \equiv y = (x \to y) \wedge (y \to x)$ —(equivalence).

Note, that $x \equiv y = 1$ only if $x = y$. Thus in this case, the value of the formula $J_1(x \equiv y)$ is 1. Otherwise the value is 0.

Proof of (C):

1. $x < y$. Then $C = ((0 \to x) \wedge 1) \vee \neg x = 1 \vee \neg 1 = 1 = x \rightarrowtail y$.
2. $x = y$. Then $C = ((1 \to x) \wedge 1) \vee \neg x = x \vee \neg x$. We have two subcases.
   (a) $x \neq 0$. Then $\neg x = 0$, and hence $C = x \vee 0 = x = x \rightarrowtail y$.
   (b) $x = 0$. Then $\neg x = 1$, and hence $C = 0 \vee 1 = 1 = x \rightarrowtail y$.
3. $x > y$. Then $C = ((0 \to x) \wedge (x \to y)) \vee \neg x = (x \to y) \vee 0 = x \to y = x \rightarrowtail y$.

■

It follows from Lemmas 0.1 and 0.2

Theorem 0.3

$$\mathcal{L}_1 = \mathcal{L}_2.$$

This theorem is important, as we shall see, for investigating algebraic properties of a logic without fixed points.

Now we can produce another non-standard model $\mathcal{M}_\Sigma$ without fixed point, which it was proposed by the author in 1983 for quite different purposes:

$$\mathcal{M}_\Sigma\langle \Sigma; \{0^-\}, \sim, \mapsto \rangle,$$

where type-order of the set Σ is $\omega + \omega^*$, that is

$$\begin{aligned}
&\Sigma = \{0^+, 1, 2, \ldots \ldots, -2, -1, 0^-\}, \\
&\sim x = -x, \\
&x \mapsto y = \begin{cases} 0^-, & \text{if } x \leq y \\ y - x, & \text{if } x > y, \end{cases} \\
&x \vee y = (x \mapsto y) \mapsto y, \\
&x \wedge y = \sim (\sim x \vee \sim y).
\end{aligned}$$

Note that matrix $\mathcal{M}_\Sigma$, in contradistinction to $\mathcal{M}^L_{\chi_0}$, is a discrete model for Łukasiewicz's infinite- valued logic t_{χ_0} which was axiomatized in the following manner:

1. $p \to (q \to p)$
2. $(p \to q) \to ((q \to r) \to (p \to r))$
3. $((p \to q) \to q) \to ((q \to p) \to p)$
4. $((\sim p \to \sim q) \to (q \to p))$.

Rules: substitution, modus ponens.

Its completeness relative to $\mathcal{M}^L_{\chi_0}$ was proved by Chang [1] (see also [2]) by introducing MV-algebras. In turn, it was proved by Vasyukov [17] that if we take Σ from $\mathcal{M}_\Sigma$ as the truth-value set in case of the definition of an evaluation in possible worlds semantics with the ternary accessibility relation for $Ł_{\chi_0}$ then such a model is correct for $Ł_{\chi_0}$.

Since $\mathcal{M}_\Sigma$ is a matrix without fixed points, operation $x \rightarrowtail y$ can be uniquely defined on the set Σ. Let us denote the new matrix by $\mathcal{M}^*_\Sigma$:

$$\mathcal{M}^*_\Sigma = \langle \Sigma; \{0^-\}, \sim x \rightarrowtail \rangle,$$

where

$$x \rightarrowtail y = \begin{cases} x, & \text{if } x = y, \text{ and } 0^+ < x < 0^- \\ x \mapsto y, & \text{otherwise.} \end{cases}$$

So in matrix $\mathcal{M}^*_\Sigma$ operations $x \mapsto y, \neg x$ and $x \Rightarrow y$ are defined analogously to $\mathcal{M}^*_{\chi_0}$ matrix.

Thus, matrices $\mathcal{M}^*_{\chi_0}$ and $\mathcal{M}^*_\Sigma$ are modelling an extension of $Ł_{\chi_0}$ by adding to it of operation $\neg x$./ This extension we denote $\mapsto Ł^*_{\chi_0}$. It would be of the interest to consider an algebraic representation of $Ł^*_{\chi_0}$ taking into account the above proved Theorem 0.3, definition of $x \Rightarrow y$, and also that $x \oplus y = \sim x \to y$.

So we have the following algebraic structure of $Ł^*_{\chi_0}$:

$\langle A, \vee, \wedge, \Rightarrow, \oplus, \sim, 0, 1 \rangle$, where
$\langle A, \vee, \wedge, \sim, 0, 1 \rangle$ de Morgan algebra, or quasi-Boolean algebra [13, p. 44],
$\langle A, \vee, \wedge, \Rightarrow, 0, 1 \rangle$Heyting algebra, or pseudo-Boolean algebra [13, p. 62],
$\langle A \vee, \wedge, \Rightarrow, \sim, 0, 1 \rangle$ symmetrical Heyting algebra [6].

Since $x \mathbin{\dot{-}} y = \sim (\sim y \Rightarrow \sim x)$ then

$$\langle A, \vee, \wedge, \Rightarrow, \dot{-}, 0, 1 \rangle$$

H-B-algebra, or double Heyting (Brouwerian) algebra, or semi-Boolean algebra [14].

Since $\neg x = x \Rightarrow 0$, and $\ulcorner x - 1 \mathbin{\dot{-}} x$, then

$\langle A, \vee, \wedge, \neg, \ulcorner, 0, 1 \rangle$ double p- algebra [8],
$\langle A, \oplus, \cdot, \sim, 0, 1 \rangle$ MV-algebra $(x \cdot y = \sim (\sim x \oplus \sim y)$, or
$\langle A, \to, \sim, 1 \rangle$ Wajsberg algebra [4],
$\langle A, *, 0, 1 \rangle$ bounded commutative BK-algebra [12] .

Each of these last three algebras, enriched with the operation $\neg x$, is adequate for $Ł^*_{\chi_0}$ in virtue of Theorem 0.3. The same holds for the algebra which has been dubbed symmetrical Heyting monoid (SHM) [7] (but in this paper the matrix logic $Ł^*_{\chi_0}$ was not described):

$$\text{SHM} = \langle A, \Rightarrow, \oplus, \sim, 0, 1 \rangle,$$

whose axiomatics are as follows:

A.1. Axioms of positive implication algebra $\langle A, \Rightarrow, 1 \rangle$ [13, p. 22].

1.1. $x \Rightarrow (y \Rightarrow x) = 1$

1.2. $(x \Rightarrow (y \to z)) \Rightarrow ((x \Rightarrow y) \Rightarrow (x \Rightarrow z)) = 1$

1.3. if $x \Rightarrow y = 1$ and $y \to x = 1$, then $x = y$

1.4. $x \Rightarrow 1 = 1$.

A.2. Axioms of Abelian monoid $\langle A, \oplus, 0 \rangle$:

2.1. $x \oplus (y \oplus z) = (x \oplus y) \oplus z$

2.2. $x \oplus y = y \oplus x$

2.3. $x \oplus 0 = x$.

A.3. Axioms for involution:

3.1. $\sim\sim x = x$

3.2. $\sim 1 = 0$

A.4. Axioms of interconnection:

4.1. $(x \Rightarrow y) \oplus z = (x \oplus z) \Rightarrow (y \oplus z)$

4.2. $\sim (x \oplus y) \oplus y = \sim (\sim x \oplus \sim y) \oplus \sim x$

4.3. $x \oplus \sim x = 1$

4.4. $x \oplus 1 = 1$.

The existence of the notion of symmetrical Heyting monoid is due to the following definition of Łukasiewicz's implication:

$$x \to y = (x \Rightarrow y) \oplus (\sim y \Rightarrow \sim x).$$

Academy of Science, Moscow, Russia

References

[1] C. C. Chang. Algebraic analysis of many-valued logic. *Transactions of the American Mathematical Society*, **88**, 467–490, 1958.

[2] C. C. Chang. A new proof of the completeness of the Łukasiewicz axioms. *Transactions of the American Mathematical Society*, **93**, 74– 80, 1959.

[3] R. Cignoli. Proper n-valued Łukasiewicz algebras as S-algebras of Łukasiewicz n-valued propositional calculi. *Studia Logica*, **41**, 3–16, 1982.

[4] J. M. Font, A. J. Rodrigues and A. Torrens. Wajsberg algebras. *Stochastica*, **3**, 5–31, 1984.

[5] K. Gödel. Zum intuitionistischen Auussagenkalkül. *Ergebnisse eines mathematischen Kolloquiums*, **4**, 40, 1931/2, published 1933.

[6] L. Iturrioz. Symmetrical Heyting algebras with operators. *Zeitschrift fü Math. Logik u. Grundl. der Mathematik*, **29**, 33–70, 1983.

[7] A. S. Karpenko and V. L. Vasykov. Symmetrical Heyting monoid: a model for infinite-valued logic of Łukasiewicz, in: *Non-classical Logics and Propositional Attitudes*, eds V. A. Smirnov *et al.*, pp. 118–124, Moscow, 1987. (In Russian)

[8] T. Katrinak. The structure of distributive double p-algebras. *Algebra Universalis*, **4**, 259–267, 1974.

[9] J. Łukasiewicz and A. Tarski. Untersuchungen üder den Aussagenkalkül. *C. R. Acad. Scie. Lett. Varsovie*, **23**, 30–50, 1930. (English translation in *Selected works of J. Łukasiewicz*, ed L. Borkowski, North Holland Publ. Co., Amsterdam, 1970.)

[10] R. McNaughton. A theorem about infinite-valued sentential logic. *The Journal of Symbolic Logic*, **16**, 1–13, 1951.

[11] Gr. Moisil. Les logiques non chrysippinnes et leurs applications. *Acta Phil. Fennica*, **16**, 137–152, 1963.

[12] D. Mundici. MV-algebras are categorically equivalent to bounded commutative BCK- algebras. *Mathematical Japonica*, **31**, 889–894, 1986.

[13] H. Rasiowa. An algebraic approach to non-classical logics.

[14] C. Rauszer. Semi-Boolean algebras and their application to intuitionistic logic with dual operations. *Fundamenta Mathematicae*, **83**, 219–29, 1974.

[15] N. Rescher. *Many-valued Logic*. McGraw-Hill, NY, 1969.

[16] J. B. Rosser and Turquette. *Many-valued Logics*. North-Holland, Amsterdam, 1952.

[17] V. L. Vasiukow. The completeness of $Ł_{\chi_0}$ under a factor-semantic. In *Non-classical Logics and its Applications*, pp. 56–68, eds V. A. Smirnov *et al.*, Moscow, 1989. (In Russian)

Part V

Analysis of Logical Proofs

15

An Extension of Gentzen's Analysis of Logical Deduction to Second-Order Logic

O. F. Serebriannikov

As far as this author knows, the existent proofs of the cut-elimination or normal form theorem make use of means that are not formalized in first-order arithmetic and in this sense cannot be considered elementary (cf. [2, 4, 5, 6]). The non-elementary proofs give rise to some kind of intellectual uneasiness because they take for granted (even if in a metalanguage) what should be proved.

I have found the proof of the cut-elimination theorem for second-order logic using a new method which I hold to be an extension of the original Gentzen's method (Cf. [7]). The proof in question may hardly be considered more complex than the proof of Gentzen's Hauptsatz for first-order logic [1].

In order to make the proof of the theorem under consideration as simple as possible I have chosen as an appropriate formal system the system of natural deduction, described and called in [4] intuitionistic theory of species. For the purpose of this paper it is convenient to add the following rule of reflexivity.

$$\frac{A}{A}.$$

Procedure of assigning. One and only one of two signs (−), (+) can be ascribed to each formula in a derivation (inserting some application of reflexivity if necessary) according to the rule:

1. the sign (−0 is ascribed to assumptions;

2. if the sign (−) (resp. (+)) is ascribed to the premise (conclusion) of reflexivity, then the sign (+) (resp. (−)) is ascribed to its conclusion (premise);

3. the sign (−) is ascribed both to the premise and conclusion of the first (second)-order ∀-elimination;

4. the sign (+) is ascribed both to the premise and conclusion of the first (second)-order ∀-introduction;

P. I. Bystrov and V. N. Sadovsky (eds.), Philosophical Logic and Logical Philosophy, 223–226.

5. the sign $(+)$ (resp. $(-)$) is ascribed to the minor (resp major) premise of the $\supset$-elimination; the conclusion of the $\supset$- elimination has the sign $(-)$;

6. the sign $(+)$ is ascribed both to the premise and conclusion of the $\supset$-introduction.

A derivation D is called assigned if in D one of the signs $(-), (+)$ is ascribed to each formula in D. In the sequel only the assigned derivations are considered. According to which of the signs $(-), (+)$ is ascribed to a derivation formula all the derivation formulas are divided in positive and negative ones in the natural way.

An application of the reflexivity in a derivation is normal or non-normal according to whether its premise is positive or negative, respectively. The non-normal application of the reflexivity is called a cut. In turn, a derivation is said to be normal, or cut-free, if it does not contain any cut.

Open assumptions of a derivation, i.e. assumptions upon which its end formula depends and its end formula itself are called terminal formulas.

The notion of a path as defined below plays an important part in the analysis of logical deduction. By a path in a derivation D from a terminal formula A to a terminal formula B in D, I understand a sequence

$$A_1, A_2, \ldots, A_n (n > 2)$$

of formulas in D such that A (graphically) coincides with A_1, A_n coincides with B and one of the following conditions holds for each $i < n$:

1. A_i is the premise (conclusion), and A_{i+2} is the conclusion (premise) of the rules: reflexivity, $\forall$-introduction, $\forall$- elimination;

2. A_i is the premise (conclusion) of the $\supset$-introduction, and A_{i+1} is the conclusion (premise) of the $\supset$-introduction;

3. A_i is the conclusion of an application of the $\supset$-introduction, A_{i+1} is an assumption closed (discharged) by this application; or vice versa;

4. A_i is the minor premise, A_{i+1} is the major premise of the $\supset$-elimination, or vice versa;

5. A_i is the major premise, A_{i+1} is the conclusion of the $\supset$-elimination; or vice versa;

6. for each clause (1)–(5) takes place the following: A_i and A_{i+2} do not proceed from one and the same formula in D.

A path in a derivation D is said to be irreducible if it does not contain two segments such that (1) they are of the same shape and, (2) their corresponding members proceed from one and the same derivation formula in D, (3) cancelling one of them yields another path in D from the same to the same terminal formula.

The length of a path is the number of its members. Now for a derivation D the measure of its non-normality is defined as the sum of the lengths of all paths in D which contain a pair of formulas proceeding from a non-normal application of reflexivity in D. Thus, zero is the measure of non-normality of a cut-free derivation.

The list of reduction presented in [4] includes the first order $\forall$- reduction, the second-order $\forall$-reduction and $\supset$-reduction. For the assigned derivations this list must be completed by the trivial reduction eliminating cuts whose conclusions are premises of (normal) applications of reflexivity.

Trivial reduction

$$\begin{array}{ccc} \bullet & & \bullet \\ \bullet & & \bullet \\ \dfrac{\dfrac{(+)A}{(-)A}}{(+)A} & \text{reduces to} & \begin{array}{c} (+)A \\ \bullet \\ \bullet \\ \bullet \end{array} \\ \bullet & & \\ \bullet & & \\ \bullet & & \end{array}$$

It is easy to see that the $\supset$-reduction when applied to any derivation of the measure k of non-normality with respect to a cut whose conclusion is the major premise of such an application of the $\supset$-elimination that there are no cuts above its minor premise yields a derivation of the measure $i < k$ with the same end formula and with the same or smaller list of open assumptions. As to the first (second) order $\forall$-reduction one can easily show that each of them transforms any derivation in a derivation of the smaller measure of non-normality with the same end formula and the same list of open assumptions. Thus for every M^2-derivation of the measure $k > 1$ one can construct an M^2-derivation of the measure $i < k$ with the same end formula and the same or smaller list of open assumptions. The latter proves the cut-elimination, or normal form, theorem for the system M^2.

Theorem. *Any M^2-derivation with the end formula F and the list of open assumptions Γ can be transformed in a cut-free M^2-derivation with the end formula F and the list of open assumptions Γ', obtained from Γ by cancelling zero or more members.*

The system C^2 of natural deduction for classical second-order logic is obtained by adjoining the following rule to M^2.

$$\frac{\begin{array}{c}[(-)E] \\ (+)\bot\end{array}}{(+)E},$$

where E is an atomic formula, $\bot$ is the abbreviation for $\forall YY$ and the signs $(-)$, $(+)$ indicate the way of assigning applications of that rule in a derivation. If in the above rule the restriction on E is cancelled by demanding that E is any formula, not identical to $\bot$, then the obtained rule can be derived in C^2.

Remark. The formulation of the cut-elimination theorem for C^2 is obtained with the aid of replacing M^2 by C^2 in the above formulation of the corresponding theory for M^2. Taking into account the new rule the relevant proof can be obtained in a similar manner.

Remark. The above elementary proof of the cut-elimination theorem together with the results of Section 17 in [6] implies an elementary consistency roof of second-order Peano arithmetic.

According to the finite point of view developed in the first volume of the famous work of Hilbert and Bernays [3], a consistency proof cannot be considered irreproachable if in it the means which go beyond the first-order arithmetic are available. The results presented in this paper may give a new stimulus to discuss the question in what form the consistency proofs of sufficiently rich formalised theories are possible.

Leningrad State University, Russia.

References

[1] G. Gentzen. Untersuchungen über daslogische Schliessen. *Mathematische Zeitschrift*, **35**, 1934.

[2] J.-Y. Girard. Une extension de l'interprétation de Gödel à l'analyse et son application à l'éliminatin des coupures dans l'analyse et la théorie de types. In *Proceedings of the second Scandinavian Logic Symposium*, Amsterdam, London 1971.

[3] D. Hilbert and P. Bernays. *Grundlagen der Mathematik 1.* Zweite Auflage, Berlin, 1968.

[4] P. Matin-Löf. Hauptsatz for theory of species. In *Proceedings of the second Scandinavian Logic Symposium*, Amsterdam, London 1971.

[5] D. Prawitz. Ideas and Results in proof theory. In *Proceedings of the second Scandinavian Logic Symposium*, Amsterdam, London 1971.

[6] G. Takeuti. *Proof Theory* Amsterdam, 1975.

[7] O. F. Serebriannikov. An elementary proof ofthe cut-elimination theorem in second orde logic. Issledovaniya po neklassicheskim logikam. Moskow 1989. (In Russian.)

16

The Sequent Variant of System R

V. M. Popov

There is a sequent formulation (calculus Gd) of one of the important and most well-motivated relevant systems—system R—suggested here. The indexing (by means of index d) of formulas occurring in sequents of the calculi GRd is stipulated by two natural (from the viewpoint of relevant logic) demands. These demands are the following:

1. it is necessary to distinguish the conjunctive and implicative connections between the members of an antecedent of a sequent;

2. it is necessary to distinguish the disjunctive and implicative- negative connections between the members of succedent of sequent.

Calculus Gd gives rather great heuristic possibilities for the search of formulas which are provable in the system R.

The language $\mathcal{L}$ of the system R is standard propositional language the alphabet of which contains propositional variables $p_1, p_2, p_3, \ldots$, binary logical connectives $\&, \vee, \supset$, unary logical connective $\neg$ and round brackets. $\mathcal{L}$-formulas will be denoted by capital letters of Latin alphabet (possibly with lower indexes). Schemes of R are the following thirteen schemes A1–A13 (see [1])

A1 $\quad A \supset A$

A2 $\quad A \supset ((A \supset B) \supset B)$

A3 $\quad (A \supset B) \supset ((B \supset C) \supset (A \supset C))$

A4 $\quad (A \supset (\supset B)) \supset (A \supset B)$

A5 $\quad A\&B \supset A$

A6 $\quad A\&B \supset B$

A7 $\quad ((A \supset B)\&(A \supset C)) \supset (A \supset (B\&C))$

A8 $\quad A \supset A \vee B$

P. I. Bystrov and V. N. Sadovsky (eds.), Philosophical Logic and Logical Philosophy, 227–234.

A9 $B \supset A \vee B$

A10 $((A \supset C)\&(B \supset C)) \supset ((A \vee B) \supset C)$

A11 $A\&(B \vee C) \supset (A\&B) \vee (A\&C)$

A12 $(A \supset \neg B) \supset (B \supset \neg A)$

A13 $\neg\neg A \supset A$

Rules of inference:

$$\frac{A \quad A \supset B}{B} \qquad\qquad \frac{A \quad B}{A\&B}$$

the notion of R-proof is usual. In order to formulate the calculus GRd it is necessary to give a number of definitions. Let each $\mathcal{L}$-formula A connect with the expression which is called d-expression of the formula A.

The definition of quasiformula:
α is a quasiformula if α is $\mathcal{L}$-formula or d- expression of the $\mathcal{L}$-formula.

By means of $\Gamma, \Delta, \Theta, \Sigma$ we denote the finite(possibly empty) sequences of formulas, and by means of Λ- the empty sequence of quasiformulas.

The definition of the unary operation d on the set of all finite (including empty) sequences of quasiformulas:[1]

1. $d(\Lambda) = \Lambda$
2. $d(\Gamma, A) = d(\Gamma), A^d$
3. $d(\Gamma, A^d) = d(\Gamma), A^d$

We are also in need of a binary operation $*$, which is defined on the set of all (including empty) finite sequences of quasiformulas:

$$\Gamma * \Delta = \begin{cases} \Gamma \text{ if } \Gamma = \Delta; \\ d(\Gamma), d(\Delta) \text{ if } \Gamma \neq \Delta \end{cases}$$

Now we are in a position to describe the sequent calculi GRd. Let any expression of the form $\Gamma \to \Delta$ where Γ and Δ are any finite sequences of quasiformulas (including the cases when one of these or even both are empty) be called indexed sequent (or simply a sequent[2]). The set of all main sequents of GRd coincides with the set of all sequents of the form $A \to A$. The following fifteen rules R1–R15 are the rules of inference of GRd.

[1] Here and everywhere below a comma between the formal expressions means the operation on concatenation of sequences; the formulas singled out are identified with one-element sequences of formulas.

[2] We follow [2] in distinguishing the notions 'sequence of formulas', 'list of formulas' and 'set of formulas'.

$$\frac{A, \Gamma \rightarrow \Delta, B}{\Gamma \rightarrow \Delta, A \supset B} \text{(R1)} \qquad \frac{\Gamma \rightarrow \Delta, A \quad B, \Sigma \rightarrow \Theta}{A \supset B, \Gamma, \Sigma \rightarrow \Delta, \Theta} \text{(R2)}$$

$$\frac{A, \Gamma \rightarrow \Delta}{\Gamma \rightarrow \Delta, \neg A} \text{(R3)} \qquad \frac{\Gamma \rightarrow \Delta, A}{\neg A, \Gamma \rightarrow \Delta} \text{(R4)}$$

Let $o_i(i = 1, 2, 3)$ be index d or the empty word, then expression A^{o_i} is d-expression of $\mathcal{L}$-formula A or $\mathcal{L}$-formula A itself.

$$\frac{\Gamma \rightarrow \Theta, A^{o_1}\Sigma \rightarrow \Theta, B^{o_1}}{\Gamma * \Sigma \rightarrow \Theta, A\&^{o_1}B} \text{(R5)}$$

where both Γ and Σ are non-empty or empty (simultaneously);

$$\frac{A^{o_1}, \Gamma \rightarrow \Delta}{A\&^{o_1}B, \Gamma \rightarrow \Delta} \text{(R6)} \qquad \frac{A^{o_1}, \Gamma \rightarrow \Delta}{B\&^{o_1}A, \Gamma \rightarrow \Delta} \text{(R7)}$$

$$\frac{\Gamma \rightarrow \Delta, A^{o_1}}{\Gamma \rightarrow \Delta, A \vee^{o_1} B} \text{(R8)} \qquad \frac{\Gamma \rightarrow \Delta, A^{o_1}}{\Gamma \rightarrow \Delta, B \vee^{o_1} A} \text{(R9)}$$

$$\frac{A^{o_1}, \Gamma \rightarrow \Theta \quad B^{o_1}, \Gamma \rightarrow \Delta}{A \vee^{o_1} B, \Gamma \rightarrow \Theta * \Delta} \text{(R10)},$$

where both Θ and Δ are non-empty or empty (simultaneously)

$$\frac{\Gamma, A^{o_1}, B^{o_2}, \Delta \rightarrow \Theta}{\Gamma, B^{o_2}, A^{o_1}, \Delta \rightarrow \Theta} box(R11) \qquad \frac{\Gamma \rightarrow \Delta, A^{o_1}, B^{o_2}, \Theta}{\Gamma \rightarrow \Delta, B^{o_2}, A^{o_1}, \Theta} \text{(R12)}$$

$$\frac{A^{o_1}, A^{o_2}, \Gamma \rightarrow \Theta}{A^{o_3}, \Gamma \rightarrow \Theta} \text{(R13)} \qquad \frac{\Delta \rightarrow \Gamma, A^{o_1}, A^{o_2}}{\Delta \rightarrow \Gamma, A^{o_3}} \text{(R14)}$$

In the rules R13 and R14 o_3 is an empty word iff at least one of the following conditions (I) or (II) is satisfied:

I o_1 is an empty word and o_2 is an empty word;

II Γ does not contain the occurrences of any d- expression.

The last rule is cut

$$\frac{\Gamma \rightarrow \Delta, A \quad A, \Sigma \rightarrow \Theta}{\Gamma, \Sigma \rightarrow \Delta, \Theta} \text{(R15)}$$

The notion of proof in GRd is a standard one.

To prove the deductive equivalence of R and GRd we use the notion of formal image of sequent which is introduced by means of the following definitions.

The definition of pure sequent:

$\Gamma \rightarrow \Delta$ is pure sequent iff neither Γ nor Δ contains any d-expression.

The definition of mapping Φ of the set of all pure sequents into the set of all $\mathcal{L}$-formulas:

$$\Phi(\Gamma \to \Delta) = \begin{cases} A_1 \supset (A_2 \supset \ldots (A_n \supset (\neg B_1 \supset (\neg B_2 \supset \ldots \\ \quad (\neg B_{n-1} \supset B_m) \ldots))) \ldots), \\ \text{if } \Gamma \to \Delta \text{ is } A_1, A_2, \ldots, A_n \to B_1, B_2, \ldots, B_m \\ (n \in \{0, 1, 2, \ldots\}, m \in \{1, 2, \ldots\}); \\ A_1 \supset (A_2 \ldots (A_{n-1} \supset \neg A_n) \ldots), \\ \text{if } |\Gamma \to \Delta \text{ is } A_1, A_2, \ldots, A_n \to \\ (n \in \{1, 2, \ldots\}); \\ \neg(p_1 \supset p_1) \text{ if } \Gamma \to \Delta \text{ is } \to \end{cases}$$

Φ correlates each pure sequent to a $\mathcal{L}$-formula.

The definition of the operation f:

f is the operation on the set of all finite sequences of quasiformulas (including the empty sequence of quasiformulas) which satisfies the following three conditions:

1f. $f(\Lambda) = \Lambda$;

2f. $f(\Gamma, A) = f(\Gamma), A$;

3f. $f(\Gamma, A^d) = A^d, f(\Gamma)$.

The definition of the regular sequence of quasiformulas: π is regular quasiformulas sequence iff no occurrence of $\mathcal{L}$- formula in π precedes any occurrence of d-expression in π.

The definition of the operation g:

g is the operation on the set of all finite regular sequences of quasiformulas (including the empty) which satisfies the following conditions:

1g. $g(\Lambda) = \Lambda$;

2g. $g(A, \Gamma) - A, \Gamma$;

3g1. $g(A^d) = A$;

3g2. $g(A_1^d, A_2^d) = A_1 \& A_2$;

3g3. $g(A_1^d, A_2^d, A_3^d) = (A_1 \& A_2) \& A_3$;

$\vdots$

4g. $g(A_1^d, \ldots, A_n, B, \Gamma) = g(A_1^d, \ldots, A_n^d), G, \Gamma$.

We substitute in this definition the symbol 'g' by the symbol 'h' and '&' by '$\vee$', one can get the definition of the operation h.

The definition of mapping F of the set of all sequents into the set of all $\mathcal{L}$-formulas:

$$F(\Gamma \to \Delta) = \Phi(g(f(\Gamma)) \to h(f(\Delta))).$$

F correlates each sequence to $\mathcal{L}$-formula (note that for the pure sequents F coincides with Φ).

The $\mathcal{L}$-formula $F(\Gamma \to \Delta)$ is a formal image of the sequent $\Gamma \to \Delta$.

Two following theorems T_1 and T_2 determine the connection between R and GRd.

T_1 If $\mathcal{L}$-formula A is provable in R, then the sequent $\to A$ is provable in GRd.

T_2 If sequent $\Gamma \to \Delta$ is provable in GRd, then $\mathcal{L}$-formula $F(\Gamma \to \Delta)$ is provable in R.

To prove the theorem T_1 it is quite enough to show that :

1. for any axiom A of the system R the sequent $\to A$ is provable in GRd;

2. if the sequents $\to A$ and $\to A \supset B$ are provable in GRd, then $\to B$ is provable too.
   If the sequents $\to A$ and $\to B$ are provable in GRd then $\to A \& B$ is provable too.

It is easy to see that GRd-proofs of substitution instances of any axiom- scheme of the system R except A11 can be constructed without using any d-expression. Detailed checking of this we leave for readers.

The GRd-proof of any substitution instance of A11 is the following:

$$
\dfrac{\dfrac{\dfrac{\dfrac{\dfrac{\dfrac{\dfrac{\dfrac{\dfrac{\dfrac{\dfrac{\dfrac{A \to A \quad B \to B}{A^d, B^d \to A\&B}\,(\mathrm{R5})}{B^d, A^d \to A\&B}\,(\mathrm{R11}) \qquad \dfrac{\dfrac{A \to A \quad C \to C}{A^d, C^d \to A\&C}\,(\mathrm{R5})}{C^d, A^d \to A\&C}\,(\mathrm{R11})}{B \vee^d C, A^d \to A\&^d B, A\&^d C}\,(\mathrm{R10})}{A\&^d(B \vee C), A^d \to A\&^d B, A\&^d C}\,(\mathrm{R7})}{A^d, A\&^d(B \vee C) \to A\&^d B, A\&^d C}\,(\mathrm{R11})}{A\&^d(B \vee C), A\&^d(B \vee C) \to A\&^d B, A\&^d C}\,(\mathrm{R6})}{A\&(B \vee C) \to A\&^d B, A\&^d C}\,(\mathrm{R13})}{A\&(B \vee C) \to A\&^d B, (A\&B) \vee^d (A\&C)}\,(\mathrm{R9})}{A\&(B \vee C) \to (A\&B) \vee^d (A\&C), A\&^d B}\,(\mathrm{R12})}{A\&(B \vee C) \to (A\&B) \vee^d (A\&C), (A\&B) \vee^d (A\&C)}\,(\mathrm{R8})}{A\&(B \vee C) \to (A\&B) \vee (A\&C)}\,(\mathrm{R14})}{\to A\&(B \vee C) \supset (A\&B) \vee (A\&C)}\,(\mathrm{R1})
$$

The proof of the proposition (2) is quite simple. Let $\dfrac{K_i}{\to A}$ and $\dfrac{K_2}{\to A \supset B}$ are GRd-proofs. Then we have the following GRd-proof of sequent $\to B$.

$$
\dfrac{\dfrac{K_1}{\to A} \qquad \dfrac{\dfrac{K_2}{\to A \supset B} \qquad \dfrac{A \to A \quad B \to B}{A \supset B, A \to B}\,(\mathrm{R2})}{A \to B}\,(\mathrm{R15})}{\to B}\,(\mathrm{R15})
$$

Let $\dfrac{K_i}{\to A}$ and $\dfrac{K_2}{\to B}$ are GRd-proofs. Then we have the following GRd-proof of sequent $\to A\&B$.

$$
\dfrac{\dfrac{K_1}{\to A} \qquad \dfrac{K_2}{\to B}}{\to A\&B}\,(\mathrm{R5})
$$

To prove the theorem T_2 it is enough to show that:

1′ for any $\mathcal{L}$-formula A $\mathcal{L}$-formula $A \supset A$ is provable in R;

2′ for any given application of single-premise rule of GRd if the formal image of premise of this application is provable in R, then the formal image of conclusion is provable in R; the same holds for any given application of two-premised rule: if the formal images of both premises are provable in R, then the formal image of conclusion is provable in R.

The proof of $(1')$ is trivial.

When proving $(2')$ it is convenient to use the notion of inference BDOC2 (see [2, pp. 52–54]) for the system R and the corresponding form of deduction theorem (see [2, pp. 69–671]).

The definition of inference BDOC2 for the system R (the notion of height of inference is defined simultaneously):

a) If A is the axiom of R, then $\overline{A}$ is inference of $\mathcal{L}$-formula A from the empty list of premises (the height of this inference is equal to 1);

b) $\mathcal{L}$-formula A is the inference of $\mathcal{L}$-formula A from the list of premises A (the height of this inference is equal to 0);

c) If α is the inference of $\mathcal{L}$-formula A from the list of premises Γ, β is the inference of $\mathcal{L}$-formula $A \supset B$ from the list of premises Δ, then $\dfrac{\alpha \quad \beta}{B}$ is inference of $\mathcal{L}$-formula B from the list of premises Γ, Δ (the height of this inference exceeds by a 1 the greater one of α and β);

d) if α is the inference of $\mathcal{L}$-formula A from the empty list of premises, β is the inference of $\mathcal{L}$-formula B from the empty list of premises, then $\dfrac{\alpha \quad \beta}{A\&B}$ is the inference of $\mathcal{L}$-formula $A\&B$ from the empty list of premises (the height of this inference exceeds by a 1 the greater one of α and β).

Now one can obtain the proof of $(2')$ by means of the deduction theorem formulated below considering the cases (which depend on the form of rule of the calculus *GRd*) without any difficulties. So we confine the consideration by formulation of Deduction theorem.

Deduction Theorem: If the inference of the $\mathcal{L}$-formula B from the list A, Γ exists, then the inference of $\mathcal{L}$-formula $A \supset B$ from the list Γ exist too.

The proof is usual by the induction on the length of formulas.

In the papers [3, 4, 5] which contain the pioneer ideas in the study of relevant logics by constructing the calculi of indexed sequents, a number of cut-free calculi were suggested. All of these formalize some 'relevant' extensions of systems E and R. The cut-elimination theorem, in all probability, does not hold for GRd, because there are serious reasons to say that the calculus obtained from *GRd* is decidable only by elimination of cut. But propositional system R, in accordance to [6] is undecidable.

Kalinin State University, Russia

References

[1] A. Anderson and N. Belnap. Entailment, Vol 1. Princeton, 1975.

[2] V. A. Smirnov. Formal deduction and logical calculi, Moscow, 1972. (In Russian)

[3] P. I. Bystrov. Embedding of relevant system E into tableaux-calculus of indexed formuli. Many valued, relevant and paraconsistent logics. *Proceedings of the Scientific Seminar in Logic of the Institute of Philosophy*, pp. 20–26, Moscow, 1984. (In Russian)

[4] P. I. Bystrov. The calculi of indexed sequents in relevant logic. Intensional logics and logical structure of theories (Abstracts of the 4th Soviet–Finnish Symposium in Logic, pp. 31–32, Tbilisi, 1985. (In Russian)

[5] P. I. Bystrov. Tableaux variants of some modal and relevant systems. *Bulletin of the Section of Logic*, **17**, Warszaw–Lódz, 1988.

[6] A. Urquhart. The undecidability of entailment and relevant implication. *Journal of Symbolic Logic*, **49**, 1-59–1073, 1984.

17

Non-standard Sequent Calculi for Modal and Relevant Logics

Peter Bystrov

1 Introduction

The search of cut-free formulations for modal, relevant, paraconsistent, etc., logics is interesting and nontrivial of itself. But it should be noted that a more fundamental goal is desirable to achieve, namely to use appropriate extensions of syntactical tools in a constructive proving of the transition from the cut-elimination theorem to the model-theoretic results. The present paper is concerned with the first as well as the second. I am trying to show that syntactical methods based on Gentzen's ideas are powerful and promising tools in studying non-classical logics.

More specifically, modified Gentzen-type sequent calculi for modal and relevant propositional logics will be constructed for each of which the cut- elimination theorem holds. Modifications in question are based on the following two key ideas: to use some global[1] rules of inference when constructing a proof tree; to make the difference between occurrences of the same formula (the same list of formulas) independent of its position in a proof.

Various techniques in doing so can be employed, for instance, special syntactical operators, indexes, scribe marks, etc. The last two mentioned will be used in the present contribution.

2 Modal Sequent Calculi with Global Rules of Inference

In formalization throughout this section $\supset$ (implication), & (conjunction), $\vee$ (disjunction) and $\neg$ (negation) will be taken as logical connectives, and $\Box$ (necessity)

[1] It can be said that a rule of inference is local if its application depends on the form of premise (premises). A rule is considered to be global if its application is regulated by the proof conditions of its premise (premises).

P. I. Bystrov and V. N. Sadovsky (eds.), Philosophical Logic and Logical Philosophy, 235–255.

as modal operator. Formulas are defined in the usual way. We use Latin letters $p, q, r, \ldots$ for propositional letters and $A, B, C, \ldots$ for formulas, and abbreviate $\neg\Box\neg$ as a $\Diamond$.

Now we will introduce modal logics Gt, G4, G4.2, G5 and GBr which are formulated in modified Gentzen-type calculi all being deductively equivalent to well known axiomatic normal modal systems T, S4, S4.2, S5 and Br (Brower's modal system), respectively.

In the following Greek capital letters $\Gamma, \Delta, \Theta, Z$ will be used for finite (possibly empty) sequences of formulas separated by commas. An expression of the form $\Gamma \rightarrow \Theta$ is called a sequent. It is assumed here a familiarity with the basic knowledge of Gentzen-type systems. The terminology used without explanations is borrowed from [7, 10, 5, 12].

An occurrence of some formula in a sequent S belonging to a proof tree D will be called a *sequent formula* and S itself will be called *D- sequent.* An ancestor of the sequent formula in a derivation is called a *parametric ancestor* if neither it nor its descendant occurs to be a side formula of a logical rule. The *rank* of a sequent formula in a derivation is the greatest number of its parametric ancestors in the same thread except those belonging to premises of permutations.

Roughly speaking GT, G4, G4.2, G5 and GBr are the formal systems obtained from a propositional fragment of calculus G (see [10]) for classical logic by adding special modal rules. To formulate these the most important notion of *path in a derivation* must be previously introduced.

Definition 2.1 A path W in a derivation D from sequent formula A to sequent formula C of D-sequent S is a shortest sequence.

$$(*) \quad B_1, B_2, \ldots, B_n \quad (n > 1),$$

such that:

1. B_1 is A;
2. for every $i (i = 1, 2, \ldots, n)$, B_i is a parametric ancestor of the main formula and B_{i+1} is a parametric ancestor of the side formula of an application of some rule in D; or B_i is a parametric ancestor of the side formula and B_{i+1} is a parametric ancestor of the main formula of an application of some rule in D; or B_i is a parametric ancestor of the antecedent and B_{i+1} is a parametric ancestor of a succedent of an occurrence of the initial sequent (axiom) in D; or B_i is a parametric ancestor of a succedent and B_{i+1} is a parametric ancestor of an antecedent of an occurrence of the initial sequent in D; or B_i is a parametric ancestor of the left (right) and B_{i+1} is a parametric ancestor of the right (left) side formula of an application of the cut in D;
3. B_n is C.

Formula B_i in a sequence $(*)$ is called a *component* of the path W in the derivation D, B_i being a *positive* (or *negative*) if it graphically coincides with a succedent (or antecedent) sequent formula of some D-sequent S, respectively. A *contrary pair* of the path W are the components A_i, A_j of W such that A_i coincides with antecedent (succedent) and A_j—with succedent (antecedent) of an occurrence of initial sequent in D. One can extend the notion 'ancestor' to paths as well, i.e. a component A_k of W is ancestor of a component A_l if A_k, A_l coincide with such sequent formulas A_k, A_l that are connected (in fixed order) by the relation 'to be an ancestor' in D. Rewriting $(*)$ in the reverse order gives the path W that will be called *inversion* of W.

We next formulate the condition for the components of such paths which contain formulas prefixed by $\Box$, i.e. formulas of the form $\Box B$.

(CON) Let W be a path and A, B are its components. Then in W (1) A precedes to B ; (2) B is positive and prefixed by $\Box$; (3) A is negative and prefixed by $\Box$; (4) if one of A, B coincides with a side formula of the cut then another does not.

Definition 2.2 A is *conjugated* with B in a path W if A is a component nearest to B in W, satisfying (CON) and conjugated with no component which precedes to B in W.

We are now in position to formulate the global modal rules for the calculi GM[2] listed above adding to G two following rule- schemes

$$\to \Box\frac{\Gamma \to \Theta, A}{\Gamma \to \Theta, \Box A}\,; \qquad \Box \to \frac{A, \Gamma \to \Theta}{\Box A, \Gamma \to \Theta}\,.$$

Let a sequent tree will be called GM-tree if all its top sequents are axioms and others are obtained by applications of the rules of GM that will be called GM-rules.

Definition 2.3 Sequent formula C of a D-sequent S in GM-tree D is T-*restricted* iff for each path W_i from any sequent formula A of S to C there is negative component of the form $\Box B$, such that (1) $\Box B$ precedes to the member F of contrary pair which is an ancestor of $\Box B$ or coincides with F; (2) no positive component prefixed by $\Box$ for which $\Box B$ is non-parametric ancestor precedes to $\Box B$; (3) $\Box B$ is conjugated with no component of W_i or its inversion W_i'.

Definition 2.4 of S4-*restricted* sequent formula results from the Definition 2.3 when dropping out clause (3) and, accordingly replacing the notation T by S4.

Definition 2.5 of S4.2-*restricted* sequent formula results from the Definition 2.4 when dropping out clause (2) and replacing notation S4 by S4.2.

Definition 2.6 Sequent formula C of a D-sequent S in GM-tree D is S5-*restricted* iff for each path W_i from any sequent formula A of S to C there is negative component prefixed by $\Box$.

Definition 2.7 of Br-*restricted* sequent formula results from the Definition 2.6 by adding the following phrase: 'and conjugated with no component of W_i or its inversion W_i'.

[2]We use GM and M as metanotations here and everywhere below. One can obtain the notation of specific calculus replacing the letter M by the notation T, S4, etc.

The Definitions 2.3–2.7 give us an important notion of formula restricted in the sense of certain modal system. Using this notion we complete the formulation of GM by definition of proof.

Definition 2.8 GM-*proof* is GM-tree D such that a side formula of each application of the rule $\rightarrow \Box$ in D is M-restricted formula.

The cut-elimination theorem holds for GM.

Theorem 2.9 *If there is GM-proof of a sequent S then GM-proof of the same sequent can be constructed without application of cut.*

Note that definitions of M-restricted formulas regulate a form of GM-tree for each specific system and, in fact, reflect an order of applications of modal rules in the proof of premises used in fixed application of the rule $\rightarrow \Box$ beginning from an axiom (axioms). The same effect can be obtained by restricting not the proof but the form of premise of $\rightarrow \Box$. For instance, one can require each application of the rule $\rightarrow \Box$ in GT-proof to be of the form

$$\frac{\Gamma \rightarrow A}{\Box\Gamma \rightarrow \Box A}$$

and so obtain the traditional sequent version of system T. This mean that proof of equivalence for GT and T can be obtained without any obstacles. The details of the proof are left for the reader.

Let an occurrence of $\Box$ in a formula be positive if this occurrence is in the scope of $2n (n = 0, 1, 2, \ldots)$ occurrences of $\neg$ and negative if it is in the scope of $2n - 1 (n = 1, 2, \ldots)$ occurrences of $\neg$, an occurrence of $\Box$ in the antecedent of implication being considered as its occurrence in the scope of $\neg$.

Definition 2.10 formula A is S4.2-modalized iff each propositional variable of A occurs in the scope of positive occurrence of $\Box$.

Definition 2.11 of S4-modalized formula results when one adds to the Definition 2.10 the following phrase: 'which is outside the scope of negative occurrence of $\Box$'.

Now we can state the following lemma for GM′ (note that in this case M′ can be replaced by S4, S4.2 or S5 only).

Lemma 2.12 *If formula A is M′-modalized then sequent $A \rightarrow \Box A$ is provable in GM′.*

Then it is quite easy to prove the following lemmas as well.

Lemma 2.13 *If formula A is provable in M′ then sequent $\rightarrow A$ is provable in GM′.*

PROOF. Proof is standard: it is easy to construct GM′-proof of the sequent $\rightarrow A$ if A is an axiom of M′, and we use the cut in the case A is obtained by means of *modus ponens* and $\rightarrow \Box$ when A is obtained by Gödel's rule. ∎

Lemma 2.14 *If* $\Gamma \to \Theta$ *is provable in GM′ then* $\varphi(\Gamma \to \Theta)$ *is provable in M′, where* $\varphi(\Gamma \to \Theta)$ *is translation (in the usual sense) of sequent* $\Gamma \to \Theta$.

PROOF. Proof can be carried out as following. Since the cut-elimination theorem holds for GM′ and $\varphi(A \to A)$ is trivially provable in M it is enough to show that for each application of a rule R in cut-free GM′- proof if translation of the premise (premises) of R is (are) provable in M′, the conclusion of R is provable too. Certainly the last can be done using Lemma 2.12 and an appropriate theorem of M′. ∎

The following theorem of equivalence for GM′ and M′ follows immediately from Lemmas 2.13 and 2.14.

Theorem 2.15 *Formula A is provable in M iff there is GM′-proof of sequent* $\to A$.

The proof of deductive equivalence for Br and BFr is a bit more complicated but it can be carried out too. In this case we previously prove intertransformation of normal GBr-proofs and GBr°-proofs where GBr° is GT with the additional sequent $A \to \Box\Diamond A$ and then show the equivalence of GBr° and Br. This observation concludes the consideration of interconnections between GM and axiomatic modal systems.

The formulations of modal logics described above was first introduced by O. F. Serebriannikov and studied in [16]. The strong cut-elimination theorem for quantified GM systems has been proved in [15]. Clearly GM-systems can be formulated with two initial modal operators $\Box$ and $\Diamond$ (see [1]) that may be useful for studying modal logics in which the equivalences $\neg\Box\neg A \equiv \Diamond A$ and $\neg\Diamond\neg A \equiv \Box A$ does not hold.

3 Indexed Sequent Calculi for Modal Logics

It turns out that the constructions of modal systems described in the preceding section are not *ad hoc* or 'purely artificial'. The notions of M-restricted sequent formulas regulating the correctness of $\to \Box$ rule applications reflects, so to say, syntactical behaviour of a modal operator in a proof. Indexes ascribed to sequent formulas or D- sequents can advantageously play the same role. To show it explicitly the alternative constructions of GM-systems will be considered in this section. All notations introduced above are supposed to be preserved each special change being pointed out.

We make use of the method relating to [9] and [19] with an appropriate modifications. At the first place these concerns with notion of index and modal rules of inference.

Index is a finite sequence of natural numbers beginning with 0 in which no number is repeated. Using letters $u, w, w_1, w_2, \ldots$ for indexes and letters $k, l, m, n, \ldots$ for natural numbers we define a binary relation R on the set of indexes as following.

Definition 3.1 uRw iff $w = u, k$ for some natural number k, and there are the extensions of R such that

1. $uR_{ref}w$ iff uRw or $u = w$;

2. $uR_{tr}w$ iff $uR_{ref}w$ or for some $n \geq 1$ there are indexes $w_1, w_2, \ldots,_n$ such that $w_1 = u, w_n = w$ and for each i $(1 \leq i \leq n)w_iRw_{i+1}$ holds;

3. $uR_{sym}w$ iff $uR_{ref}w$ or $wR_{ref}u$;

4. $uR_{eq}w$ iff $uR_{sym}w$ and $uR_{tr}w$.

It is obvious that R is a relation of adding a new natural number to an index, (1)–(3) specifies R as a relation which is reflexive, reflexive and transitive, symmetric and nontransitive while (4)—as a relation of equivalence for any pair of indexes.

Indices u will be called a *subindex* of index w iff $uR_{tr}w$.

Rank (or *length*) r of index w is a total quantity of natural numbers different from 0 contained in w. Expression $r(w) - = 2$ means 'the rank of index w equals two'.

An expression of the form $A^w, \neg^w A, A\&^w B, A \vee^w B, , A \supset^w B$ are considered to be indexed formulas (I-formulas); in the case $r(w) = 0$ we simply have formulas in the usual sense, i.e. A^0 means the same as A.

An expression $\Gamma \rightarrow \Theta$ will be called an *indexed sequent* (I-sequent), whereΓ, Θ are finite (possibly empty) sets of I-formulas. A number of occurrences of one and the same I-formula and order of sequent I-formulas in a I-sequent will not be taken into consideration when comparing an occurrence of I-sequents in a proof.

An I-sequent S is *pure sequent* iff $r(w) = 0$ is maximal rank for each occurrence of I-formula in S. It means that a pure sequent is just the same thing as a sequent in the usual sense. In general case of *rank of I-sequent* S is greatest of all ranks of the indexes ascribed to I- formulas in S.

Now we formulate modal calculi of I-sequents on the base of the calculus G^I which consists of the initial sequent

$$\Gamma_1, A^w, \Gamma \rightarrow \Theta_1, A^w, \Theta_2,$$

where $\Gamma_1, \Gamma_2, \Theta_1, \Theta_2$ are finite (possibly empty) sets of I-formulas, and the following rules of inference

$$\frac{A^w, \Gamma \rightarrow \Theta, A \supset^w B, B}{\Gamma \rightarrow \Theta, A \supset^w B} \qquad \frac{A \supset^w B, \Gamma \rightarrow \Theta, A^w, \quad B^w, A \supset^w B, \Gamma \rightarrow \Theta}{A \supset^w B, \Gamma \rightarrow \Theta}$$

$$\frac{A^w, \Gamma \rightarrow \Theta, \neg^w A}{\Gamma \rightarrow \Theta, \neg^w A} \qquad \frac{\neg^w A, \Gamma \rightarrow \Theta, A^w}{\neg^w A, \Gamma \rightarrow \Theta}$$

$$\frac{A^w, B^w, A\&^w B, \Gamma \rightarrow \Theta}{A\&^w B, \Gamma \rightarrow \Theta} \qquad \frac{\Gamma \rightarrow \Theta, A\&^w B, A^w \quad \Gamma \rightarrow \Theta, A\&^w B, B^w}{\Gamma \rightarrow \Theta, A\&^w B}$$

$$\frac{A^w, A \vee^w B, \Gamma \to \Theta \quad B^w, A \vee^w B, \Gamma \to \Theta}{A \vee^w B, \Gamma \to \Theta} \quad \frac{\Gamma \to \Theta, A \vee^w B, A^w, B^w}{\Gamma \to \Theta, A \vee^w B}$$

Modal calculus GT^I is formulated by adding to the GT^I the rules intended to govern insertion of □ into succedent and antecedent of I-sequent.

$$\to \Box^I \frac{\Gamma \to \Theta, \Box A^u, A^{u,k}}{\Gamma \to \Theta, \Box A^u},$$

where k is a number which does not occur in indexes ascribed to any I- formula's occurrence in conclusion;

$$\Box \to^I \frac{A^{w_1}, \Box A^w, \Gamma \to \Theta}{\Box^w A, \Gamma \to \Theta},$$

where w_1 is an index ascribed to an occurrence of I-formula in conclusion such that $wR_{ref}w_1$.

To obtain the calculi $GS4^I$, GBr^I and $GS5^I$ it is enough to replace in the condition for indexes in formulation of the rule $\Box \to^I$ relation R_{ref} by R_{tr}, R_{sym} and R_{eq} accordingly.

Naturally GM^I-proofs will be all such GM^I trees of I- sequents which are constructed from axioms by means of the rules formulated above.

The main peculiarities of DM^I-proofs are an absence of the structural rules and retaining a main formula in premises of logical rules. Moreover, for each correct GM^I-proof an application of initial I- sequent is supposed to be fixed, i.e. a certain analysis of a proof is given that enables us to avoid vagueness in grasping what the contrary pair (the occurrences of one and the same I-formula on the left and right from arrow) of an GM^I-axiom is.

Example of $GS5^I$-*proof*:

$$\cfrac{\cfrac{\cfrac{\cfrac{\cfrac{\cfrac{A^{0,2}, \Box^{0,1} A, \neg^0 \Box A, \to \neg\Box A \supset^0 \Box\neg\Box A, \Box^0 \neg\Box A, \neg^{0,1} \Box A, \Box^0 A, A^{0,2}}{\Box^{0,1} A, \neg^0 \Box A \to \neg\Box A \supset^0 \Box\neg\Box A, \Box^0 \neg\Box A, \neg^{0,1} \Box A, \Box^0 A, A^{0,2}}}{\neg^{0,1} \Box A \to \neg\Box A \supset^0 \Box\neg\Box A, \Box^0 \neg\Box A, \neg^{0,1} \Box A, \Box^0 A, A^{0,2}}}{\neg^0 \Box A \to \neg\Box A \supset^0 \Box\neg\Box A, \Box^0 \neg\Box A, \neg^{0,1} \Box A, \Box^0 A}}{\neg^0 \Box A \to \neg\Box A \supset^0 \Box\neg\Box A, \Box^0 \neg\Box A, \neg^{0,1} J\Box A}}{\neg^0 \Box A \to \neg\Box A \supset^0 \Box\neg\Box A, \Box^0 \neg\Box A}}{\to \neg\Box A \supset^0 \Box\neg\Box A}$$

It is easy to check that this GM^I -tree is none of the GT^I, $GS4^I$, GBr^I-proofs.

Remark. It is comparatively easy to show by induction on the number of DI-sequents (where DI-sequent is considered to be a specific occurrence of I-sequent in a proof) contained in a GM^I- proof that for any w the following rules

$$\frac{\Gamma \rightarrow \Theta}{\Gamma \rightarrow \Theta, A^w} \quad \text{and} \quad \frac{\Gamma \rightarrow \Theta}{A^w, \Gamma \rightarrow \Theta}$$

are derivable in GM^I.

We continue the consideration comparing GM^I-calculi with M and GM by means of the following definition of *image* $\varphi(S)$ of I-sequent S in which an induction on the rank r of I-sequent S is used.

Definition 3.2 Let S be an I-sequent

$$A_1^u, \ldots, A_n^w \rightarrow B_1^u, \ldots, B_m^w$$

and r is a rank of S.

1. If $r = 0$ then $\varphi(S)$ is a disjunction of the form

$$\neg A_1 \vee \ldots \vee \neg A_n \vee B_1 \vee \ldots \vee B_m;$$

2. if $r > 0$ hen $\varphi(S) = \varphi(S_1)$, where S_1 is I-sequent

$$\rightarrow \neg^u A_1, \ldots, \neg^w A_n, B_1^u, \quad B_m^w,$$

   rank of S_1 is $R(w) = l$, and $\varphi(S_1)$ is

$$\varphi^*(\rightarrow \neg^u A_1, \ldots, B_1^u, \ldots, \Box^{w_1}(A_n \vee B_m)),$$

   where $r(w_1) = l - 1$, disjunction prefixed by $\Box$ is constructed from the I-formulas with the index w in accordance with the rule $\rightarrow \Box^I$ and φ^* is already defined by the clause (1).

Uniformly speaking the Definition 3.2 presents the following procedure of constructing $\varphi(S)$ for I-sequent S:

1. if S is pure I-sequent, i.e. its rank equals to 0, then we construct its image as usual;
2. if rank of S is > 0, then we
   (a) transform S into S_1 by transferring all antecedent I-formulas of S, each being prefixed by $\neg$, to its succedent and retaining every index without changes;
   (b) pick out in groups by permutations in S_1 the I-formulas with the same indexes of the rank r and reform each of such groups into disjunction prefixed by $\Box$;

(c) delete the indexes ascribed to members of disjunction and ascribe to the modal prefix of disjunction a new index in accordance with the rule $\to \Box^{\mathrm{I}}$. Thus a sequent with a rank $r-1$ is obtained for which an image may already have been defined.

Using the Definition 3.2 it can be proved that each GM^{I}- proof of I-sequent S can be transformed not M-proof of $\varphi(S)$ (see details in [4]). There are, however, other ways to obtain deductive equivalence of GM^{I} and M when rather complicated direct proof of the cut-elimination theorem for GM^{I} is not previously required. We will show this by using lemmas concerned with a certain relationship between GM and GM^{I} instead of proving the equivalence of GM^{I} and M directly.

Definition 3.3 The M-*indexation* of GM-proof D is an index's ascribing procedure for every DI-sequent of D such that

1. index 0 is ascribed to each I-formula in the end sequent of D;
2. the same index is ascribed to a parametric formula in conclusion and premise (premises) of a GM-rule;
3. the same index is ascribed to a side formula (formulas) and a main formula of a non-modal GM-rule;
4. index u, k is ascribed to a side formula of $\to \Box$ if index u is ascribed to main formula of this rule, k being a new natural number which has no occurrences in the conclusion;
5. index w_1 is ascribed to a side formula of $\Box \to$ such that wRw_1, where w is an index ascribed to main formula and w_1 is subindex of some index ascribed to a formula occurring in the conclusion.

M-indexation of GM-proof D is considered to be correct if, as a result of application of the clauses (1)–(5), the same index is ascribed to both members of initial sequent for each of its specific occurrences in D.

Using this definition, then, we are able to move from GM-proof to GM^{I}-proof. Note, at first, that the following lemma holds.

Lemma 3.4 *Every cut-free GM-proof D in which each antecedent formula of the form $\Box A$ is a descendant of a main formula of some application of the rule $\Box \to$ admits a correct M-indexation such that main and side formulas both have index 0 if there are no applications of the rule $\to \Box$ below the application of $\Box \to$ in question.*

Proof can be performed by induction on a number of D-sequents in a GM- proof in accordance with clauses of Definition 3.3 without difficulty. Note that Lemma 3.4 permits simply to use in M-indexation of a $\to \Box$-free proof the relation R_{ref} only. It immediately follows from the above Remark on the derivability of the weakening in GM^{I} and Lemma 3.4 that the next lemma can easily be verified.

Lemma 3.5 *Every cut-free GM-proof of pure sequent S can be transformed into GM^{I}-proof of this sequent.*

As it turns out, the restrictions similar to those for applications of $\rightarrow \Box$ in GM-proofs can be extended to GM^{I}-proofs with fixed analysis of GM^{I}-axioms' applications which makes indexes to be excessive in a modal sequent tree.

Definition 3.6 Sequent formula A^w of DI-sequent S is M-*restricted* iff w differs from each index ascribed to other sequent formulas in S.

Extending M-restriction into GM^{I}-proofs we shall then state

Lemma 3.7 *Every GM^{I}-proof of pure sequent S can be transformed into GM-proof of S^* by deleting all indexes, each formula of S^* being M- restricted if the corresponding sequent formula of S is M-restricted.*

Proof is by induction on the number of DI-sequents in initial GM^{I}-proof.

This means that GM and GM^{I}- are deductively equivalent.

Theorem 3.8 *There is GM^{I}-proof of pure sequent S iff GM-proof of S can be constructed.*

Clearly, the theorem follows from Lemmas 3.5 and 3.7. So we immediately obtain the counterpart of the cut-elimination theorem for GM^{I}-.

Theorem 3.9 *Whatever GM^{I}-proof of the pure sequents $A, \Gamma \rightarrow \Theta$ and $\Delta \rightarrow Z, A$ may be GM^{I}-proof of the pure sequent $\Gamma, \Delta \rightarrow \Theta, Z$ can be constructed.*

It is sufficient then, to get the following theorem.

Theorem 3.10 *GM is deductively equivalent to M.*

The last follows from Theorem 3.9 and result from the previous section concerned with the equivalence of GM and M.

Note that the definition of R_{eq} allows us to present GS5^{I}- just as in Kanger-style considering indexes simply as natural numbers but not as sequences of these. Thus Theorem 3.9 holds for Kanger's system S5* from [9].

One can observe that formulations of GM and GM^{I} presented above are more complicated in comparison with standard sequent counterparts of the same modal systems. This is true. But it must be noticed that GM gives us the cut-free variants of the systems S5 and Br preserving subformula property for sequent proofs. As to GM^{I} these give us a good constructive decision procedure for modal systems in question as well as for some extensions of the system S4 (see for example, [3, 4]). Furthermore it can be shown that GM^{I} are useful for constructive proof of completeness of modal systems with respect to Kripke-style models.

4 Indexed Sequent Calculi and Models for Modalities

In this section we shall describe the techniques for transition from modal sequent calculi to semantic models. Note, at first, that considering GM^I–calculi it is possible to avoid I-formulas and to use an indexation of sequents only. This way, as will be shown below, leads us directly to such modal calculi for which Kripke's tableaux are simply an abbreviation of proof's notations. In other words, our goal now is to carry out a direct transition from GM^I to Kripke's tableaux, i.e. to models for M.

At first we shall formulate an auxiliary calculus of sets of I-sequents as useful tools for going from trees of I-sequents to tableaux constructions.

Definition 4.1 A *set of I-sequents* is an expression of the form $S_1; S_2; \ldots; S_n$, where $S_i (i = 1, 2, \ldots, n)$ is an I-sequent and $n \geq 0$.

Calculus BM^I- of sets of I-sequents includes an axiom-set of I- sequents $S_1; \ldots;$ S_k such that for each $i (i = 1, \ldots, k)$ S_i is GM^I-axiom and $k > 0$; and the following two schemes of rules:

$$B1\frac{\mathbb{U}; \mathbb{S}'; \mathbb{T}}{\mathbb{U}; \mathbb{S}; \mathbb{T}} \qquad B1\frac{\mathbb{U}; \mathbb{S}'; \mathbb{T}; \mathbb{U}; \mathbb{S}''; \mathbb{T}}{\mathbb{U}; \mathbb{S}; \mathbb{T}}$$

where $\mathbb{U}$ and $\mathbb{T}$ are sets of I-sequents; in B1 $\mathbb{S}'$ is a premise and $\mathbb{S}$ is a conclusion of some one-premise rule of GM^I; in B2 $\mathbb{S}'$ and $\mathbb{S}''$ are premises (left and right accordingly) and $\mathbb{S}$ is the conclusion of a two-premise GM^I-rule.

BM^I-proof of a I-sequent S is a proof of the set of I-sequents containing S only.

Lemma 4.2 *Every GM^I-proof of I-sequent S can be transformed into BM^I-proof of S*

This can be verified by induction on a number of threads in an initial GM^I-proof.

Lemma 4.3 *Every BM^I-proof of a set $S_1; S_2; \ldots; S_n$ can be transformed into exactly n GM^I-proofs of I-sequents $S_1, S_2, \ldots, S_n$ accordingly.*

The proof is by induction on a number of D-sets in an initial BM^I-proof, where D-set is considered to be an occurrence of a set of I- sequent in BM^I-proof.

It is obvious that the equivalence of GM^I- and BM^I immediately follows from Lemmas 4.2 and 4.3.

Now we introduce a notion of *index-sequent* which is to be distinguished from the notion of I-sequent. Let $\Gamma \to \Theta$ be a pure sequent as it was defined in the previous section. Then the expression of the form $\Gamma \to^w \Theta$ where w is an index will be an *index-sequent* of the rank $r(w)$.

Definition 4.4 An *abseqeunt* is an expression of the form $S_1 \cdot S_2 \cdot \ldots \cdot S_m$, where for each $i (i = 1, 2, \ldots, m) S_i$ is an index-sequent and $m \geq 0$.

A *set of absequents* is defined by analogy with a set of I-sequents, namely it is an expression of the form

$$(\mathrm{SA})\mathbb{W}_1;\mathbb{W}_2;\ldots;\mathbb{W}_n,$$

where for each $i(i = 1, 2, \ldots, n)\mathbb{W}_i$ is an absequent.

Let (SA) be called an axiom of the calculus KM^{I} of sets of absequents if $n > 0$ and each member $\mathbb{W}_i$ of (SA) contains an abseqeunt with a contrary pair. Then it is convenient to describe the rules of KM^{I} by comparison with BM^{I}-rules. We say that every application of a BM^{I}-rule is transformed in application of a KM^{I}-rule by means of the following replacement of I-sequents to corresponding absequents. Let S be an I-sequent. The (i) fix all groups of I-formulas with the same indexes in S; (ii) for each such group construct an index-sequent transferring indexes from formulas to arrows and deleting others I-formulas of S; construct an abseqeunt S' from index- sequents obtained by the previous steps.

As a result of the steps (i)–(ii) we have absequent S' corresponding to I- sequent S. The notion of an I-sequent corresponding to absequent can be defined in a similar way.

Examples. For the purpose of abbreviation we suppose $\alpha = \neg\Box\neg A \&^0 \Box(\Box\neg A \vee B)$, $\beta = \neg^0\Box\neg A$ and $\gamma = \Box^0(\Box\neg A \vee B)$.

$\mathrm{BS5}^{\mathrm{I}}$-proof of the sequent $\alpha \to \Box B$:

$$\cfrac{\cfrac{\cfrac{\cfrac{\cfrac{\cfrac{\cfrac{\begin{array}{c}\neg^{01}A, \Box^{02}\neg A, \Box\neg A^{02} \vee B, \beta^0, \gamma^0, \alpha^0 \to \Box^0 B, \Box^0\neg A B^{02}, \neg^{01}A;\\ B^{02}, \Box\neg A^{02} \vee B, \beta^0, \gamma^0, \alpha^0 \to \Box^0 B, \Box^0\neg A, B^{02}\end{array}}{\begin{array}{c}\Box^{02}\neg A, \Box\neg A^{02} \vee B, \beta^0, \gamma^0, \alpha^0 \to \Box^0 B, \Box^0\neg A, B^{02}, \neg^{01}A;\\ B^{02}, \Box\neg A^{02} \vee B, \beta^0, \gamma^0, \alpha^0 \to \Box^0 B, \Box^0\neg A, B^{02}\end{array}}}{\Box\neg A^{02} \vee B, \beta^0, \gamma^0, \alpha^0 \to \Box^0 B, \Box^0\neg A, B^{02}, \neg^{01}A}}{\beta^0, \gamma^0, \alpha^0 \to \Box^0 B, \Box\neg A, B^{02}, \neg^{01}A}}{\beta^0, \gamma^0, \alpha^0 \to \Box^0 B, \Box^0 A, \neg^{01}A}}{\beta^0, \gamma^0, \alpha^0 \to \Box^0 B, \Box^0\neg A}}{\beta^0, \gamma^0, \alpha^0 \to \Box^0 B}}{\alpha^0 \to \Box^0 B}$$

One can easily transform it into the following $\mathrm{KS5}^{\mathrm{I}}$-proof:

$$\dfrac{\dfrac{\dfrac{\dfrac{\dfrac{\dfrac{\dfrac{\Box\neg A, \Box\neg A \vee B \rightarrow^{02} B \cdot \beta\gamma\alpha \rightarrow^{0} \Box\neg A, \Box B \cdot \neg A \rightarrow^{01} \neg A; \quad B \rightarrow^{02} B \cdot \beta, \gamma\alpha \rightarrow^{0} \Box\neg A, \Box B \cdot \rightarrow^{01} \neg A}{\Box\neg A, \Box\neg A \vee B \rightarrow^{02} B \cdot \beta, \gamma, \alpha \rightarrow^{0} \Box\neg A, \Box B \cdot \rightarrow^{01} \neg A; \quad B \rightarrow^{02} B \cdot \beta, \gamma, \alpha \rightarrow^{0} \Box\neg A, \Box G \cdot \rightarrow^{01} \neg A}}{\Box\neg A \vee B \rightarrow^{02} B \cdot \beta, \gamma, \alpha \rightarrow^{0} \Box\neg A, \Box B \cdot \rightarrow^{01} \neg A}}{\beta, \gamma, \alpha \rightarrow^{0} \Box\neg A, \Box B \cdot \rightarrow^{01} \neg A \cdot \rightarrow^{02} B}}{\beta, \gamma, \alpha \rightarrow^{0} \Box\neg A, \Box B \cdot \rightarrow^{01} \neg A}}{\beta, \gamma, \alpha \rightarrow^{0} \Box B, \Box\neg A}}{\beta, \gamma, \alpha \rightarrow^{0} \Box B}}{\alpha \rightarrow^{0} \Box B}$$

Now in order to do the last step towards the semantical level we show that every KM^{I}-proof can be recorded in a closed Kripke tableau construction.

As usual KM^{I}-construction for a pure sequent $\Gamma \rightarrow \Theta$ is a system of alternative sets of tableaux which is begun by the formulas in Γ on the left and those in Θ on the right of the initial tableau (t_0-tableau). Then we apply standard Beth's rules for classical logical constants and the following two rules for $\Box$.

($\Box_1$) When $\Box A$ occurs on the left of tableau t_w belonging to an alternative set K write A on the left of each tableau t_{w_1} such that t_{w_1} belongs to K and $t_w R_M t_{w_1}$. (M can be replaced here by *ref, tr, sym* or *eq*).

($\Box_r$) When $\Box A$ occurs on the right of tableau t_u belonging to an alternative set K construct a new auxiliary tableau $t_{u,k'}$ write on the right of $t_{u,k}$ and add it to K.

Using the usual notion of a closed tableaux construction one can easily verify that KM^{I}-construction for pure sequent S closes iff there is a KM^{I}-proof of S. It is quite trivial because, in fact, KM^{I}-constructions do the job just as a set of absequents in a KM^{I}-proof.

It should be noted, firstly, that we preserve an indexation of tableaux in KM^{I}-constructions. It is done to simplify the definition of KM^{I}-construction and to show a natural character of proposed transition from sequents to models but it is not necessary in general. Secondly, one can obtain:

Theorem 4.5 *If there are KM^{I}-proofs of the pure sequents $A, \Gamma \rightarrow \Theta$ and $\Delta \rightarrow Z, A$ then KM^{I}-proof of the pure sequent $\Delta, \Gamma \rightarrow \Theta, Z$ can be constructed.*

The theorem immediately follows from Theorem 3.9 and Lemmas 4.2 and 4.3.

Now we have a counterpart of the cut-elimination theorem for tableau system and so, certainly, verify that if there are closed tableaux constructions for $\rightarrow A$ and $\rightarrow A \supset B$ then there is a closed tableau construction for B too (see Kripke's remark about the proof of completeness in the beginning of Section 4.1 of [12]). Thus there is no need to prove supplementary the equivalence of tableaux and models. We can verify *modus ponens* and rule of necessitation as well as all M-axioms iff it is possible to get a corresponding KM^{I}- construction.

Next and the last step is obvious—to associate an occurrence of A in the left column of a tableau t_w with the truth in model: $v(A, w) = T$; and in the right column—with the falsity: $v(A, w) = F$. Of course, our wRw_1 means the same as HRH_1 in terms of standard possible world semantics.

In fact availability of GM^I-proof of the sequent $\to A$ with subformula property implies that there is a closed KM^I- construction for $\to A$, therefore a countermodel for $\varphi(\to A)$ does not exist and formula A *is* valid. And if A is valid then KM^I-construction for A is closed and so by means of BM^I and GM^I we go to the provability of A in M. Thus we have the method of carrying out $\vdash_M A$ iff $\models MA$ which is mechanically applicable and effective at least for a specific family of modal logics for which a cut-free sequent counterparts can be constructed.

5 Cut-free Relevant Logic

It would be nice to establish something like that techniques and results from the previous section could be extended to relevant logics. Indeed the problem of such extensions may be represented as known two-faced one: on the syntactical level—to construct a cut-free sequent version of a relevant system; in semantics—to use a binary accessibility relation when constructing models for relevant systems.[3]

However fruitful the ternary accessibility relations are considered to be in the case of relevance logics I cannot agree with that 'with the success of the ternary approach the binary semantics became to appear old-fashioned and special' (see [6, p. 163]). If alethic modal logic is dubious from some philosophical or non-formal semantics viewpoint the relevance one is dubious by three times as much. If we deal with logical systems, say, in the limits of the first-order formal language, not with such abstract entities as structures, fields, toposes, etc. the ternary approach is forced and intricated case of binary one rather than the last is old-fashioned and special case of the first. I believe that there are some more strict formal reasons for this suggestion concerned with specific relevance logics but its consideration falls out from the limits of present paper.

Note, only, that ternary relation was used to provide a way of rejecting such 'irrelevant' formulas as $T \supset A$ and $A \supset T$, (where T denotes a tautology), particularly the $A \supset (A \supset A)$. In general, it is not single and unique say, to preserve the relevancy in some specific sense since the most crushing grief in relevant logics, in my opinion, is not connected with implication but results from presence of the distributive law between & and $\vee$ and so called structural principles: weakening, contraction and permutation. For instance, for positive contractionless subsystems of T (ticket entailment), E and R one can obtain completeness and decidabil-

[3] A possibility to use the method similar to constructing GM^I and BM^I- calculi for some relevant logics was considered in [2, 3]. I take this opportunity to correct an inexactitude in the last. On p. 101 was intended that $\mathbb{F}^*$*-proof of* α *is a closed tableau which starts with* A^0 *on the left and* B^0 *on the right when* α *is of the form* $A \supset B$*, and with* α *on the right otherwise*. The italicised sentence was inadvertently omitted.

ity as well for its non-distributive fragments (see [8, 13, 14]). The problem is that contraction and/or distributive law in the presence of other rather minimal logical principles will collapse the cut-elimination theorem and makes decidability at least highly problematic (and according to [20] unattainable at all).

The attempt to overcome these difficulties is presented here. Namely the cut-free relevant system will be constructed. Of course, in the present case $\supset$ means relevant implication so far as $\rightarrow$ is convenient to retain for sequents.

We shall distinguish two kinds of occurrences of one and the same formula in a sequent later on: marked occurrence ($*$-occurrence) and pure (usual) one. If A^* occurs in Γ or Θ n times, we say that there are n $*$-occurrences of A in the sequent $\Gamma \rightarrow \Theta$. A sequence of formulas Γ is considered to be *pure* iff it contains no $*$-occurrences of any formula, *weak* iff it contains $*$-occurrences only and *mixed* iff it contains both kinds of formulas occurrences.

So we shall use $\Gamma, \Theta, \Delta, Z$ to denote arbitrary (pure, weak or mixed) finite sequences of formulas. Accordingly, a sequent $\Gamma \rightarrow \Theta$ is pure iff both Γ, Θ contain no $*$-occurrences, weak iff only one of Γ, Θ contain $*$- occurrences only, and mixed iff Γ or Θ contain both kinds of formula occurrences.

Note that for any one-premise sequent rule of inference there is just one occurrence of a formula sequence as antecedent or succedent of premise sequent while in the scheme of two-premise rule, say,

$$\frac{\Gamma \rightarrow \Theta, A \quad B, \Gamma \rightarrow \Theta}{A \supset B, \Gamma \rightarrow \Theta}$$

there are two occurrences of Γ and two occurrences of Θ above the line.

Let us make a distinction between two kinds of occurrences of sequences (as antecedents or succedents of sequents) in the premises of some rule—*identical* and *$*$-similar*. Suppose then Γ_1 and Γ_2 to be the occurrences of sequences in the left and right premise accordingly.

Definition 5.1 Γ_1 and Θ_2 are identical iff these are pure or weak and indistinguishable graphically, i.e. contain the same number equally ordered occurrences of the same formulas (in symbols $\Gamma_1 = \Gamma_2$).

Definition 5.2 Non-empty Γ_1 and Γ_2 are $*$-similar iff these are mixed and (1) if there are any k $*$-occurrences of $A_1, \ldots, A_m$ in Γ_1 then there are exactly k pure occurrences of the same formulas in Γ_2; and (2) if there are l $*$-occurrences of $B_1, \ldots, B_n$ in Γ_2 then there exactly l pure occurrences of the same formulas in Γ_1. (In symbols $\Gamma_1 * \Gamma_2$). $A_i (i = 1, \ldots, m)$ and $B_j (j = 1, \ldots, n)$ will be called a *common irrelevant member of* Γ_1 *and* Γ_2.

It is obvious that $*$-similar sequences will become simply identical when one deletes all marks or, quite the reverse, ascribe $*$ to all pure occurrences of formulas in every sequence.

Now we formulate the relevant sequent system GR* as follows.

Axiom: $A \rightarrow A$

Logical Rules:

$$Ir\,\frac{A,\Gamma\to\Theta,B}{\Gamma\to\Theta,A\supset B},\quad Il\,\frac{\Gamma\to\Theta,AB,\Delta\to Z}{A\supset B,\Gamma\to\Theta},$$

$$Nr\,\frac{A,\Gamma\to\Theta}{\Gamma\to\Theta,\neg A},\qquad Nl\,\frac{\Gamma\to\Theta,A}{\neg A,\Gamma\to\Theta},$$

where Γ,Θ,Δ,Z are pure sequents.

Let $\langle*\rangle$ or/and $\langle*\rangle^{\circ}$ denote/s $*$ or be empty expression/s and only one of the expressions $\langle*\rangle'$, $\langle*\rangle''$ denote $*$ while another is empty in inference schemes below.

$$Kl1\,\frac{B^{\langle*\rangle},\Gamma\to\Theta}{A\&^{\langle*\rangle}B,\Gamma\to\Theta};\qquad K12\,\frac{A^{\langle*\rangle},\Gamma\to\Theta}{A\&^{\langle*\rangle}B,\Gamma\to\Theta};$$

$$Dr1\,\frac{\Gamma\to\Theta,A^{\langle*\rangle}}{\Gamma\to\Theta,A\vee^{\langle*\rangle}B};\qquad Dr2\,\frac{\Gamma\to\Theta,B^{\langle*\rangle}}{\Gamma\to\Theta,A\vee^{\langle*\rangle}B}.$$

$$Kr1\,\frac{\Gamma_1\to\Theta_1,A\Gamma_2\to\Theta_2,B}{\Gamma^{\langle*\rangle}\to\Theta^{\langle*\rangle},A\&B},\quad Dl1\,\frac{\Gamma_1,\to\Theta_1B,\Gamma_2\to\Theta_2}{A\vee B,\Gamma^{\langle*\rangle}\to\Theta^{\langle*\rangle}}.$$

where (i) if $\Gamma_1=\Gamma_2$, then $\Gamma_1=\Gamma_2=\Gamma^{\langle*\rangle}$, i.e. if Γ_1 and Γ_2 are identical then $\langle*\rangle$ is empty expression; the same holds for Θ,Θ_1,Θ_2; (ii) if $\Theta_1*\Theta_2$, then $\Theta^{\langle*\rangle}$ is a result of ascribing $*$ to all common irrelevant members of Θ_1 and Θ_2; the same holds for Γ,Γ_1 and Γ_2.

$$Kr2\frac{\Gamma_1\to\Theta_1,A^{\langle*\rangle'}\Gamma_2\to\Theta_2,B^{\langle*\rangle''}}{\Gamma^{\langle*\rangle}\to\Theta^{\langle*\rangle},A\&B},\quad Dl2\frac{A^{\langle*\rangle'},\Gamma_1\to\Theta_1B^{\langle*\rangle''},\Gamma_2\to\Theta_2}{A\vee B,\Gamma^{\langle*\rangle}\to\Theta^{\langle*\rangle}},$$

where (1) $\Theta_1*\Theta_2$ or Θ_1,Θ_2 are both weak and $\Theta_1=\Theta_2$ in Kr2; the same holds for Γ_1 and Γ_2 in D12; (2) the clauses (i), (ii) of the condition for Kr1 and Dl1 holds in any case.

Structural rules:

$$Wr\frac{\Gamma\to\Theta}{A^*,\Gamma\to\Theta},\quad Wl\frac{\Gamma\to\Theta}{\Gamma\to\Theta,A^*},$$

where Γ in Wr and Θ in Wl are non-empty.

$$Pr\frac{\Gamma\to\Theta,A^{\langle*\rangle},B^{\langle*\rangle^{\circ}},Z}{\Gamma\to\Theta,B^{\langle*\rangle^{\circ}},A^{\langle*\rangle},Z};\qquad Pl\frac{\Gamma,A^{\langle*\rangle},B^{\langle*\rangle^{\circ}},A\to\Theta}{\Gamma,B^{\langle*\rangle^{\circ}},A^{\langle*\rangle},\Delta\to\Theta}.$$

$$Cr\frac{\Gamma \to \Theta, A, A^{\langle * \rangle}}{\Gamma \to \Theta, A}, \qquad Cl\frac{A^{\langle * \rangle}, A, \Gamma \to \Theta}{A, \Gamma \to \Theta}.$$

$$\text{cut}\,\frac{\Gamma \to \Theta, A \quad A, \Delta \to Z}{\Gamma\Delta \to \Theta Z},$$

where all sequents are pure, and $\Theta(\Delta)$ do not contain the main formula of Cr (CL)-application with non-empty $\langle * \rangle$.

The notion of GR*-proof is usual for Gentzen-type calculi. To be sure, GR*-rules listed above, excluding cut and rules for $\supset$ and $\neg$ are metaschemes of the rules of inference. One can get a specific application of GM*-rule from these taking into account the kind of sequent used as premises. For instance, every application of GR*-rule is standard (classical) one if we consider the pure sequents only. Note further that a conclusion of weakening is a necessary mixed sequent. Moreover, two premise rules for & and $\vee$ as well as contradiction are in a power to make changes with marks, replacing a pure occurrence of some formula by $*$-occurrence of the same formula or vice versa, besides doing its 'standard inferential job'.

Definition 5.3 An (application of) contraction in GR*-proof is *regular* iff its side formulas are pure and one of these is main formula of an application of some logical rule.

To reflect the fact that in constructing a cut-free GR*-proof one can avoid non-regular contraction, suffice it to verify

Lemma 5.4 *Every cut-free GR*-proof of the pure sequent S can be transformed into such GR*-proof of S which contains regular contractions only.*

This can be proved by induction on the height of initial GR*-proof where height is considered to be total number of rules' applications in GR*-proof in question.

Before moving to the cut-elimination theorem let us consider the following rule of inference.

$$* - \text{mix}\,\frac{\Gamma \to \Theta, A \quad A, \Delta \to Z}{\Gamma, \Delta_{*A} \to \Theta_{*A}, Z},$$

where Γ, Z are pure, Δ, Θ possibly contain $*$- occurrences of formula A only in the proofs of both premises and $\Delta_{*A}\,\Theta_{*A}$ are obtained from Δ, Θ accordingly by deleting some $*$-occurrences of formula A.

Lemma 5.5 *Every GR*-proof of pure sequent S containing applications of the cut can be transformed into GR* proof of S containing applications of the $*$-mix only.*

Proof is trivial since the cut is a specific case of the $*$-mix and every application of the $*$-mix can be easily transformed in an application of the cut by previous contractions and permutations.

Lemma 5.6 ($*$-mix-elimination lemma). *If a pure sequent is provable in GR* then there is a proof free of applications of the rule $*$-mix for that sequent in GR*.*

As usual, the proof of this lemma involves a triple induction, the main induction on the number of applications of $*$-mix. The proof can be carried out by the method similar to that proposed by Professor V. A. Smirnov in [17] for the 'absolute' relevant predicate calculus SLA. Since there is no possibility of going into the details of the proof we confine ourselves to the following remark. In some cases the proof of $*$- mix-elimination lemma differs from the standard of Smirnov's proof. Namely, when the side formula of $*$-mix in question is in the same time the main formula of contraction with pure formulas Lemma 5.4 must be used to do the induction step in the induction rank.

By the Lemma 5.6 and earlier established deductive equivalence of $*$-mix and cut, by Lemma 5.5, we have the following theorem

Theorem 5.7 (Cut-elimination theorem). *If a pure sequent is provable in GR* then this sequent is provable without application of cut in GR*.*

Now we are in a position to show that well known propositional relevant system R is a subsystem of GR*.

Theorem 5.8 *If formula A is provable in R then sequent $\rightarrow A$ is provable in GR*.*

To prove this it is quite enough to verify:

1. for each axiom A of R, $\rightarrow A$ is provable in GR*;
2. if the sequents $\rightarrow A$ and $\rightarrow B$ are provable in GR*, then the sequent $\rightarrow A\&B$ is provable in GR* as well; if the sequents $\rightarrow A$ and $\rightarrow A \supset B$ are provable in GR* then there is a proof of the sequent $\rightarrow B$ in GR*.

Of course it can be freely done. Details are left for the reader.

It would be interesting in a supreme degree to prove that GR* is a subsystem of R too. Unfortunately I am forced to consider it to be an open question today and is supposed to be solved in paper which is in preparation. This contribution comes to an end by describing the image φ of sequent used in GR*. A mixed sequent $\Gamma \rightarrow \Theta$ is considered to be *regular* iff no pure occurrence of a formula in Γ and Θ precedes any $*$-occurrence of a formula. Note that we can deal with regular mixed sequents only since every mixed sequent can be reduced to regular form by means of permutations. Then image $\varphi(\Gamma \rightarrow \Theta)$ of the sequent $\Gamma \rightarrow \Theta$ is defined as follows:

1. $\Gamma \rightarrow \Theta$ is pure sequent.
   (a) $\varphi(\Gamma \rightarrow \Theta) = A_1 \supset (A_2 \supset \ldots (A_m \supset (\neg B_1 \supset (\neg B_2 \supset \ldots (\neg B_{n-1} \supset B_n)\ldots)))\ldots)$, if $\Gamma \rightarrow \Theta$ is $A_1, A_2, \ldots, A_m \rightarrow B_1, B_2, \ldots, B_n$, where $n > 0$ and $m \geq 0$.
   (b) $\varphi(\Gamma \rightarrow \Theta) = A_1 \supset (A_2 \supset \ldots (A_{n-1} \supset \neg A_n)\ldots)$, if $\Gamma \rightarrow \Theta$ is $A_1, A_2, \ldots, A_n \rightarrow$, where $n > 0$.
2. $\Gamma \rightarrow \Theta$ is weak sequent.

(a) $\varphi(\Gamma \to \Theta) = A_1 \supset (A_2 \supset \ldots (A_m \supset B_1 \vee B_2 \vee \ldots \vee B_n) \ldots)$, if $\Gamma \to \Theta$ is $A_1, A_2, \ldots, A_m \to B_1^*, B_2^*, \ldots, B_n^*$, where $m > 0$ and $n > 0$;

(b) $\varphi(\Gamma \to \Theta) A_1 \& A_2 \& \ldots \& A_m \supset (\neg B_1 \supset \ldots (\neg B_{m-1} \supset B_m) \ldots)$, if $\Gamma \to \Theta$ is $A_1^*, A_2^*, \ldots, A_m^* \to B_1, B_2, \ldots, B_n$, where $m > 0$ and $n > 0$.

3. $\Gamma \to \Theta$ is mixed sequent.

(a) $\varphi(\Gamma \to \Theta) = (A_1 \& \ldots \& A_m \& B_1 \& \ldots \& B_k) \supset (C_1 \vee \ldots \vee C_n \vee D_1 \vee \ldots \vee D_l)$ if $\Gamma \to \Theta$ is $A_1^*, \ldots, A_m^*, B_1, \ldots, B_k \to C_1^*, \ldots, C_n^*, D_1, \ldots, D_l$, where $m, n, k, l > 0$ or $n, l > 0$ while $m, k \geq 0$.

(b) $\varphi(\Gamma \to \Theta) = A_1 \& \ldots \& A_m \supset (B_1 \supset \ldots (B_k \supset C \vee \ldots \vee C_n) \ldots)$, if $\Gamma \to \Theta$ is $A_1^*, \ldots, A_m^*, B_1, \ldots, B_k \to C_1^{\langle * \rangle}, \ldots, C_n^{\langle * \rangle}$, where $n > 0, m, k > 0$;

(c) $\varphi(\Gamma \to \Theta) = A_1 \& \ldots \& A_m \supset (\neg D_1 \supset \ldots (\neg D_l \supset C_1 \vee \ldots \vee C_n) \ldots)$, if $\Gamma \to \Theta$ is $A_1^{\langle * \rangle}, \ldots A_m^{\langle * \rangle} \to C_1^*, \ldots, C_n^*, D_1, \ldots, D_l$, where $m > 0, n, l > 0$.

(d) $\varphi(\Gamma \to \Theta) = A_1 \& \ldots \& A_m \supset (B_1 \supset \ldots (B_{k-1} \supset \neg B_k) \ldots)$, if $\Gamma \to \Theta$ is $A_1^*, \ldots, A_m^*, B_1, \ldots, B_k \to$, where $m > 0, k > 0$.

Thus we have enough tools now to investigate whether GR*-rules (of course, excluding the cut) are derivable in R in the following sense: if image (images) of the premise (premises) of the GR*-rule is (are) provable in R then the image of the conclusion of this rule is provable in R as well.

Acknowledgements

Finally, grateful acknowledgement is expressed to Professor V. A. Smirnov, Professor O. F. Serbriannikov and Dr. Vladimir Popov for the profound remarks that gave a good impetus for the present contribution.

Russian Academy of Sciences, Moscow, Russia

References

[1] P. I. Bystrov. Normalisation of Deductions in Modal Logic, Leningrad State University, Candidate Dissertation, Lenningrad, 1981. (In Russian)

[2] P. I. Bystrov. The calculi of indexed sequents in relevant logic. *Intensional Logics and Logical Structure of Theories. Abstract of the IV Finnish-Soviet Symposium in Logic*, Tbilisi, pp. 31–32, 1985. (In Russian)

[3] P. I. Bystrov. Tableaux variants of some modal and relevant logics. *Bulletin of the Section of Logic*, **17**, 92–103, 1988.

[4] P. I. Bystrov. Syntactical analysis of the extensions of S4. *Syntactical and Semantical Analysis of Non-extensional Logics*, Nauka, Moscow, pp. 290– 305, 1989. (In Russian)

[5] H. B. Curry. *Foundations of Mathematical Logic*, New York, 1963.

[6] J. M. Dunn. A Kripke-style semantics for R-mingle using a binary accessibility relation. *Studia Logica*, **XXXV**, 163–172, 1976.

[7] G. Gentzen. Investigations into Logical Deductions. *The Collected Papers of Gerhard Gentzen*, M. Szabo, ed. North Holland, Amsterdam, 1969.

[8] S. Giambrone. TW_+ and RW_+ are decidable. *Journal of Philosophical Logic*, **14**, 235–254, 1985.

[9] S. Kanger. *Provability in Logic*, Stockholm, 1957.

[10] S. K. Kleene. Permutability of inferences in Gentzen's calculi. *LK and LJ Memor. Americ. Math. Soci.*, **10**, 1–26, 1952.

[11] S. K. Kleene. *Introduction to Metamathematics*, Princeton, NJ, 1952.

[12] S. A. Kripke. Semantical analysis of modal logic I. Normal modal propositional calculi, *Zeitschrift für Mathematishce Logik und Grundlagen der Mathematik*, **9**, 67–96, 1963.

[13] H. Ono and Y. Komori. Logics without the contraction rule. *Journal of Symbolic Logic*, **50**, 169–201, 1985.

[14] V. M. Popov. On decidability of relevant logic RAO. *Modal and Intensional Logics*, pp. 115–119, Moscow, 1978. (In Russian)

[15] O. F. Serebriannikov. Some generalisations of the normal form theorem in quantified modal logic. *Modal and Intensional Logics and its Applications to the Problems of Methodology of Sciences*, pp. 88–99, Nauka, Moscow, 1984. (In Russian)

[16] O. F. Serebriannikov and P. I. Bystrov. On the criteria of quality of logical deductions in modal logic. *Logic and Philosophical Categories*, pp. 97–112, Leningrad, 1982. (In Russian)

[17] V. A. Smirnov. Formal Deduction and Logical Calculi, Nauka, Moscow, 1972. (In Russian)

[18] V. A. Smirnov. Sequent variants of the logical systems with strong and relevant implication. *Theory of Logical Deduction*, Moscow, 1974. (In Russian)

[19] K. Schütte. Vollstandige system modaler und intuitionistucher logik. *Ergebinsse der Mathematik und ihrer Grenzgebiete*, **42**, 1968.

[20] A. Urquhart. The undecidability of entailment and relevant implication. *Journal of Symbolic Logic*, **49**, 1059–1073, 1984.

18

Axiomatic Rejection for Classical Propositional Logic

Arata Ishimoto

1 Introduction

Axiomatic rejection is a method to recursively enumerate all the formulas not provable in the given formal system by way of a recursive set of axioms and rules, not necessarily finite.

In modern times, axiomatic rejection or rejection in general could be traced back to those detailed studies of classical propositional logic and syllogistic by Łukasiewicz and other members of the Lvov–Warsaw School which began to take place as early as at the beginning of the 1920s. (For this, refer to Wolenski [18].) The research thus started culminated in Slupecki [14] and the celebrated syllogistic monograph [9] by Łukasiewicz, the latter of which has made his contributions available to the wider circles of logicians. (In this connection, Slupecki–Bryll–Wybraniec–Skardowski [15] and Staszek [17] should also be consulted.)

Nevertheless, Łukasiewicz and Slupecki did not attend to the Gentzen-structure underlying the axiomatic rejection for classical propositional logic and Aristotelian syllogistic, although it was employed more or less implicitly.

Now, the purpose of this paper is to make full use of the said Gentzen-structure in the form of the tableau method in connection with the method of axiomatic rejection confining ourselves to classical propositional logic. (The axiomatic rejection to be developed in what follows goes back to a number of works (in Japanese) such as Ishimoto [6] and Inoué [4]. These studies are also related to Gentzen- type formulations of some syllogistic systems like Kobayashi–Ishimoto [8], Inoué–Ishimoto–Kobayashi [5] and Kanai [7].)

2 Syntactic Preliminaries

We shall begin this section with the introduction of the notion of positive and negative parts of a formula due to Schütte [10, 11, 12]. It could be dispensed with. But

P. I. Bystrov and V. N. Sadovsky (eds.), Philosophical Logic and Logical Philosophy, 257–270.

the notion will have the effect of simplifying the subsequent development of the proposed theory.

Definition 2.1 The positive and negative parts of a formula A are defined only as follows:

2.11 A is a positive part of A,

2.12 If $B \vee C$ is a positive part of A, then B and C are positive parts of A,

2.13 If $\sim B$ is a positive part of A , then B is a negative part of A,

2.14 If $\sim B$ is a negative part of A, then B is a positive part of A.

The (well-formed) formulas of our classical propositional logic to be designated as CP are defined in the well known way in terms of a countably infinite collection of propositional variables (atomic formulas) and two logical symbols, namely, $\vee$ (disjunction) as well as $\sim$ (negation) along with some technical symbols. (Other logical symbols are defined, if necessary, in terms of these two logical symbols. Outermost parentheses will always be suppressed.) As has been used already, such letters as $A, B, \ldots$ will be employed as meta-logical variables for formulas as above defined. All the symbols thus introduced will be used only meta-logically in what follows.

The notion of positive and negative parts of a formula will be exemplified as follows:

$$\begin{aligned}
&F[A_+] = A, \\
&F[A_+] = \sim\sim\sim\sim (B \vee A), \\
&F[A_+] == \sim\sim\sim\sim\sim\sim (\sim\sim A \vee B), \\
&G[A_-] = \sim A, \\
&G[A_-] = \sim\sim (B \vee \sim\sim\sim A), \\
&G[A_-] = \sim\sim\sim\sim\sim A,
\end{aligned}$$

where $F[A_+](G[A_-])$ signifies that A has a specific occurrence in $F[A_+](G[A_-])$ as its positive (negative) part. (This notation again originates with Schütte.) Thus, in the above examples, A and B are different formulas so that the occurrences of A there are specified.

Such expressions as $F[A_+, B_-], G[A_-, B_+, C_-]$ and the like are analogously understood with the *proviso* that the specified formulas do not overlap with each other.

Here, it is emphasized, some subformulas of a formula are neither a positive nor a negative part thereof. For example, the A and B in

$$\begin{aligned}
&\sim (B \vee A), \\
&\sim (\sim A \vee B),
\end{aligned}$$

are such formulas.

Thus, a minimal positive (negative) part of a formula is defined to be a positive (negative) part of the formula which does not contain any positive or negative parts of the formula other than itself. As is easily seen, atomic formulas, i.e. propositional variables could be a minimal positive part of a formula, while formulas of the form $A \vee B$ could also be a minimal negative part of a formula along with atomic formulas.

We shall next introduce the operation of deleting a formula from another, where the former occurs in the latter as its positive or negative part. (This again originates with Schütte.)

Given $F[A_+](G[A_-])$, the result of deleting A therefrom is denoted by $F[\ _+](G[\ _-])$, which may be the empty expression.

Definition 2.2 The deletion of A from $F[A_+](G[A_-])$ is defined only as follows:

2.21 If $F[A_+] = A$, then $F[\ _+] =$ the empty expression,

2.22 If $F[A_+] = F_1[A \vee B_+]$ (or $F_1[B \vee A_+]$), then $F[\ _+] = F_1[B_+]$,

2.23 If $G[A_-] = F[\sim A_+]$, then $G[\ _-] = F[\ _+]$,

2.24 If $F[A_+] = G[\sim A_-]$, then $F[\ _+] = G[\ _-]$.

The deletion is obviously defined by induction on the number of the procedures 2.11–2.14 employed for specifying the formula to be deleted as a positive or negative part of the given formula. It is not difficult to see that the result of deletion is a (well-formed) formula or the empty expression.

The operation of deletion will be exemplified as follows:
For $F[A_+] = A$,
$F[\ _+] =$ the empty expression,
For $F[A_+] = G[\sim A_-] = F_1[\sim\sim A_+] = F_2[\sim\sim A \vee B_+] = \sim\sim A \vee B$,
$F[\ _+] = G[\ _-] = F_1[\ _+] = F_2[B_+] = B$,
For $G[A_-] = F[\sim A_+] = \sim A$,
$G[\ _-] = F[\ _+] =$ the empty expression,
For $G[A_-] = F_1[\sim A_+] = F_2[B \vee \sim A_+] = F_3[(B \vee \sim A) \vee C_+] = (B \vee \sim A) \vee C$,
$G[\ _-] = F_1[\ _+] = F_2[B_+] = F_3[B \vee C_+] = B \vee C$,
where A, B and C are different from each other.

3 Axioms and Rules

In the present section, we shall introduce not only the axioms and rules for provability in CP, i.e. classical propositional logic, but also those corresponding to axiomatic rejection in their Hilbert-type formulation.

Since our CP is an A-N system, we can make resort to such an axiom system as proposed by Łukasiewicz for the development of its Hilbert-type version with three axioms and two rules, namely, detachment and uniform substitution (for propositional variables).

As for the axioms and rules for axiomatic rejection, those due to Slupecki [14] and Łukasiewicz [9] will be employed.

Thus, the axiom is any fixed propositional variable specified for the purpose, which we wish to state, say, as:

3.11 $\dashv D$,

with D being a propositional variable, while the rules are the following two, namely:

3.21 $\dashv B, \vdash A \supset B \Rightarrow \dashv A$,

3.22 $\dashv A \Rightarrow \dashv B$,

where in 3.22 A is obtained from B by uniform substitution for propositional variables. (Here, after Härtig [2], $\dashv A$ means that A is axiomatically rejected by way of the axiom and rules as above described.)

For reference, an example will be given of axiomatic rejection:

(1)	$\vdash D \vee \sim\sim D \vee D. \supset D$	tautology,
(2)	$\dashv D$	3.11,
(3)	$\dashv D \vee \sim\sim D \vee D$	(2), (1), 3.21,
(4)	$\dashv A \vee \sim B \vee C$	(3), 3.22,

where A, B and C are propositional variables different from each other, and (3) obtains from (4) by substituting $D, \sim D$ and D, respectively for A, B and C. Here, in (1), (2) and (4) the disjuncts are understood to be associated appropriately, and this will continue to be our understanding likewise in the sequel.

4 Tableau Method

In this section, we shall develop the tableau method for the proposed CP not only with respect to provability, but also with respect to rejection. The tableau method has been applied to a variety of logical systems, classical, intuitionistic and modal, both propositional and predicate. (For the tableau method, refer, among others, to Smullyan [16] and Schütte [11].)

The tableau method or a tableau is defined in terms of reduction rules, which are in our case only one, namely:

$$\vee\text{-}\quad \frac{G[A \vee B_{-}]}{G[A \vee B_{-}] \vee \sim A \quad | \quad G[A \vee B_{-}] \vee \sim B.}$$

(This is a branching rule.)

The rule is sufficient to give rise to all the instances of tautology. The presence of quasi-principal formulas will facilitate the definition of Hintikka formulas to be given presently.

We shall prove for reference a number of theses of CP by the proposed tableau method.

4.11 $\bigvee_- \dfrac{\sim (A \vee B) \vee (B \vee A)}{(\sim (A \vee B) \vee (B \vee A)) \vee \sim A \;\mid\; (\sim A \vee B) \vee (B \vee A)) \vee \sim B,}$

where every branch of this tableau ends with a formula of the form $F[A_+, A_-]$. Such a branch is closed, and a tableau, every branch of which is closed, is said to be closed. A closed tableau, which is given rise to by a formula, constitutes a proof of the formula.

4.12 $\bigvee_- \dfrac{\sim (\sim A \vee B) \vee (\sim\sim B \vee \sim A)}{S \vee \sim\sim A \;\mid\; S \vee \sim B,}$

where S is the given formula.

4.13 $A \vee \sim A,$

which is already of the form $F[A_+, A_-]$, and provable by a tableau consisting of only one formula, namely 4.13.

Before proceeding to the fundamental theorem for the tableau method, we need the notion of Hintikka formulas which originates with Hintikka [3].

Definition 4.1 A Hintikka formula A is a formula which satisfies the following conditions:

4.21 A is not of the form $F[B_+, B_-]$,

4.22 If $B \vee C$ occurs in A as its negative part, then B or C occurs there as its negative part.

The following are the examples of Hintikka formulas:

$$A \vee B \vee \sim C,$$
$$B \vee \sim (B \vee A) \vee C \vee \sim A,$$
$$\sim (A \vee B) \vee \sim (C \vee \sim\sim\sim (A \vee B)) \vee \sim\sim (A \vee B) \vee \sim B,$$

where all the meta-logical variables involved are propositional variables different from each other.

Lemma 4.2 *Every Hintikka formula has at least one occurrence of an atomic formula, i.e. propositional variable either as its positive or negative part.*

The Lemma is easily proved by *reductio ad absurdum*.

We are now in a position to state and prove the fundamental theorem.

Theorem 4.3 (Fundamental theorem). *Given a formula, by reducing it by way of the reduction rule, namely,* $\vee_-$, *there obtains in a finite number of steps a tableau, each branch of which is closed, or, at least, one branch of which ends with a Hintikka formula.*

Hereby, a branch is extended until we come across a formula of the form $F[A_+, A_-]$ *obtaining a closed branch, or there do not remain in the same branch any fresh formulas which could be employed as a principal formula of a* $\vee_-$ *application.*

Moreover, if a closed tableau is obtained at all by a reduction as above specified, a closed tableau is again forthcoming by a reduction which might be different from the given one with respect to the choice of principal formulas and the order of their selection in the same branch.

A tableau constructed as above specified is called a normal tableau. The tableaux presented in this section are all normal.

The Theorem is forthcoming if we remember that there occurs only a finite number of subformulas which could be used as a principal formula of the reduction rule. This is to the effect that there take place only a finite number of $\vee_-$ applications in whatever way we extend branches adhering to the restriction as specified in the Theorem.

With a view to proving that a branch, which is not closed, ends with a Hintikka formula, let us suppose, if possible, that the formula, with which the branch ends, is not a Hintikka formula. Then, the formula would contain the occurrence of a formula of the form $A \vee B$ as its negative part without A or B occurring there as its negative part. This, however, is against the requirement that there should not remain any new formula which could be employed as the principal formula of $\vee_-$ application in extending a branch.

The last part of the theorem is left to later sections.

Lemma 4.4

$$\vdash F[A_+] \equiv .F[\ _+] \vee A,$$
$$\vdash G[A_-] \equiv .G[\ _-] \vee \sim A,$$

where dots are used for sparing parentheses.

The lemma is proved by induction on the number of the procedures 2.11–2.14 applied to specify the A as a positive (negative) part of $F[A_+](G[A_-])$.

The basis:

$$\vdash F[A_+] \equiv .\ \text{the empty expression}\ \vee A. \equiv A,$$

where the disjunction of the empty expression with a formula is identified with the formula.

Induction steps:

$\vdash F[A_+] \equiv F_1[A \vee B_+]$ (or $F_1[B \vee A_+]$) $\equiv .F_1[\ _+] \vee (A \vee B)$ (or $F_1[\ _+] \vee (B \vee A)$) (induction hypothesis).
$\equiv .(F_1[\ _+] \vee B) \vee A. \equiv .F_1[B] \vee A$ (induction hypothesis).
$\equiv .F[\ _+] \vee A,$
$\vdash G[A_-] \equiv F[\sim A_+] \equiv .F[\ _+] \vee \sim A$ (induction hypothesis).
$\equiv .G[\ _-] \vee \sim A,$
$\vdash F[A_+] \equiv G[\sim A \sim] \equiv .G[\ _-] \vee \sim\sim A$ (induction hypothesis).
$\equiv .G[\ _-] \vee A. \equiv .F[\ _-] \vee A.$

Lemma 4.5 *If A is provable by the tableau method for CP, then $\vdash A$, i.e. A is a thesis of its Hilbert-typer version.*

This is proved by the length of the tableau.

In view of Lemma 4.4, we have,

$$\vdash F[A_+, A_-] \equiv .F[\ _+,\ _-] \vee (A \vee \sim A),$$

which takes care of the basis.

The induction step is dealt with by the following theorems of CP (in its Hilbert-type formulation) and Lemma 4.4:

$$\begin{aligned} &\vdash G[A \vee B_-] \equiv .G[\ _-] \vee \sim (A \vee B). \\ &\equiv .(G[\ _-] \vee \sim (A \vee B) \vee \sim A) \wedge (G[\ _-] \vee \sim (A \vee B) \vee \sim B). \\ &\equiv .(G[A \vee B_-] \vee \sim A) \wedge (G[A \vee B_-] \vee \sim B). \end{aligned}$$

Lemma 4.6 *Given a Hintikka formula A, every positive part of A is axiomatically rejected.*

With a view to proving this Lemma, $D(\sim D)$ is substituted for every propositional variable occurring in A as its positive (negative) part, where D is a propositional variable as specified in 3.11. Such a propositional variable is certainly available in view of Lemma 4.2. If $B(C)$ is a positive (negative) part of A, the formula obtained as a result of such a substitution is denoted by $B'(C')$.

We, then, have,

$$\begin{aligned} &\vdash B' \supset D, \\ &\vdash \sim C' \supset D, \end{aligned}$$

which are proved simultaneously by induction on the length of B and C.

The basis does not present any difficulties, since

$$\begin{aligned} &\vdash D \supset D, \\ &\vdash \sim\sim D \supset D, \end{aligned}$$

are tautologies.

For taking care of induction steps, suppose that B, which is a positive part of A, is of the form $B_1 \vee B_2$.

By induction hypothesis, we have,

$$\begin{aligned} &\vdash B_1' \supset D, \\ &\vdash B_2' \supset D, \end{aligned}$$

which together yield,

$$\vdash B_1' \vee B_2'. \supset D,$$

namely,

$$\vdash B' \supset D.$$

Next, suppose that B, which is a positive part of A, is of the form $\sim B_1$ with B being a negative part of A.

By induction hypothesis, we have,

$$\vdash\sim B_1' \supset D,$$

which is nothing but,

$$\vdash B' \supset D.$$

Let C be a negative part of A having the form $C_1 \vee C_2$. Since A is a Hintikka formula, C_1 or C_2, say, C_1 constitutes a negative part of A.

By induction hypothesis, we have,

$$\vdash\sim C_1' \supset D,$$

which along with

$$\vdash\sim (C_1' \vee C_2') \supset\sim C_1',$$

gives rise to

$$\vdash\sim C' \supset D$$

as requested. The case that C_2' occurs in A as its negative part is taken care of analogously.

Lastly, if C, which is a negative part of A, is of the form $\sim C_1$, with C_1 being a positive part of A, we have by induction hypothesis,

$$\vdash C_1' \supset D,$$

from which follows,

$$\vdash\sim\sim C_1' \supset D,$$

namely,

$$\vdash\sim C' \supset D.$$

From this obtains that B' is axomatically rejected by 3.11, $\vdash B' \supset D$ and 3.21. This in turn verifies that B is also axiomatically rejected by way of the rule 3.22.

This completes the proof of Lemma 4.6.

Corollary 4.7 (Inoué). *Every Hintikka formula is axiomatically rejected.*

Since a Hintikka formula constitutes itself a positive part of the Hintikka formula, it is axiomatically rejected by Lemma 4.6 just proved.

Lemma 4.8 *Every formula, which belongs to a branch of a tableau ending with a Hintikka formula, implies the lower wones.*

This is proved by induction on the length of the branch.

The basis is straightforward.

With a view to verifying the induction step, let us assume that $G[A \vee B_-] \vee \sim A$ is a formula belonging to the branch which leads to the Hintikka formula.

By induction hypothesis, $G[A \vee B_-] \vee \sim A$ implies those located below. From this follows that $G[A \vee B_-]$ also implies the lower ones in view of Lemma 4.4.

Corollary 4.9 *Every formula, which belongs to a branch of a tableau ending with a Hintikka formula, is axiomatically rejected.*

Straightforward by Lemma 4.8 and th erule 3.21.

5 Elimination Theorem

Elimination theorem for the proposed tableau method will be proved in this section on the basis of some preparatory theorems and the results obtained in the preceding sections.

Theorem 5.1 (Dichotomy theorem). *For every A, either* $\vdash A$ *or* $\dashv A$.

For proving the Theorem, let us suppose that it is not the case that $\vdash A$, namely A is not a thesis of the Hilbert-type CP. Then, there is not any tableau which proves A (by the tableau method). In fact, the assumption to the contrary would give rise to the thesishood of A in the Hilbert-type formulation of CP, namely, $\vdash A$ by Lemma 4.5. In view of Theorem 4.3 (Fundamental theorem), this is to the effect that any (normal) tableau obtained by reducing A ends with a Hintikka formula. Then, Corollary 4.9 is resorted to, and $\dashv A$ is forthcoming, since A constitutes one of the formulas belonging to the branch ending with the Hintikka formula.

Theorem 5.2 (Contradiction theorem). *It is never the case that* $\vdash A$ *and* $\dashv A$ *at the same time.*

For proving this Theorem, it is sufficient to demonstrate that $\dashv A$ gives rise to not $\vdash A$, namely, A is not a thesis of the Hilbert- type CP. This is proved in turn by induction on the length of the axiomatic rejection for A.

The basis is taken care of by showing that not $\vdash D$, namely, D is not a theorem of the Hilbert-type CP, where D is the propositional variable as specified in 3.11. As well-known, any propositional variable is not provable in CP in its Hilbert-type version. This is reduced to the consistency of the CP, and has been demonstrated in a variety of ways.

For taking care of induction steps, suppose, $\dashv A$ is obtained from $\dashv B$ and $\vdash A \supset B$ by the rule 3.21 for axiomatic rejection. By induction hypothesis, we have derived from $\dashv B$ not $\vdash B$, i.e. B is not provable in the Hilbert-type CP.

With a view to obtaining not $\vdash A$ from $\dashv A$, let us assume, if possible, to the contrary, namely $\vdash A$, which would, then, give rise to $\vdash B$ by $\vdash A \supset B$ against the induction hypothesis that not $\vdash B$, which has been obtained from $\dashv B$. Not $\vdash A$ is, thus forthcoming, namely, A is not a thesis of the Hilbert-type CP.

For treating the induction step with respect to 3.22, i.e. the second rule for axiomatic rejection, let us suppose that $\dashv A$ and A is forthcoming from B by uniform substitution for propositional variables. Further, by induction hypothesis, $\dashv A$ has lead to not $\vdash A$.

Now, assume, if possible, $\vdash B$, which, then, would give rise to $\vdash A$, since uniform substitution is a rule for the Hilbert-type CP. But, this is against the induction hypothesis that not $\vdash A$ which is obtainable from $\dashv A$. From this follows that not $\vdash B$, namely, B is not provable in the Hilbert-type version of CP.

This completes the proof of Theorem 5.2.

Before turning to the proof of normal form theorem, we wish to settle the assertion last stated in Theorem 4.3, namely, fundamental theorem to the effect that every reduction uniquely results in a finite number of steps either in a closed tableau or in a tableau not closed. In whatever way we extend the branch as far as the requirements as given in the Theorem are observed. (Such a tableau is normal whether it be closed or not.)

For proving this statement, let us assume, if possible, that we could obtain a tableau, which is not closed, by reducing a formula A by a series of the applications of the reduction rule different from the one responsible for a closed tableau with respect to the choice of principle formulas and the order of their selection in the same branch.

If this were the case, the tableau, which obtains as a result of this new reduction (of A) different from the old one, would contain a branch which ends with a Hintikka formula. This, it is remarked, is forthcoming by Theorem 4.3 without referring to the assertion we are now proving.

In view of Corollary 4.9, we would, then, obtain $\dashv A$, and by Theorem 5.2 (Contradiction theorem), this is not possible if we remember that we already have $\vdash A$ by Lemma 4.5.

This so-to-speak uniqueness result of reductions could also be proved by way of the permutability theorem in the sense of Kleene (S. C. Kleene, Permutibility of inferences in Gentzen's calculi KL and LJ, *Memoirs of the Amer. Math. Soc*, **10**), or model-theoretically.

Normal form theorem or cut-elimination theorem, which dates back to Gentzen, is in our case a theorem of the tableau method version of CP to the effect cut rule is eliminable.

Cut rule is a rule applied in the proposed formalism in the following form:

$$\text{cut}\,\frac{F[A_-]\quad G[A_-]}{F[A\,]\vee G[A\,],}$$

where we proceed from up to down, and A is called the cut formula of the cut application.

Theorem 5.3 (Normal form theorem). *If $F[A_+]$ and $G[A_-]$ are both provable in the tableau method version of CP, then we have $F[\ _+]\vee G[\ _-]$ in the same version of CP.*

For proving this Theorem suppose, if possible, that $F[\ _+]\vee G[\ _-]$ is not provable by the tableau method for CP, the thesishood there of both $F[A_+]$ and $G[A_-]$ notwithstanding. By Theorem 4.3 (Fundamental theorem), there would, then, be a branch ending with a Hintikka formula. In view of Lemmas 4.8 and 4.6, this would give rise to $\dashv F[\ _+]\vee G[\ _-]$.

Now, we have by Lemma 4.4,

$$\begin{aligned}&\vdash F[A_+]\wedge G[A_-].\equiv.(F[\ _+]\vee A)\wedge(G[\ _-]\vee\sim A).\\&\supset.(F[\ _+]\vee G[\ _-]\vee A)\wedge(F[\ _+]\vee G[\ _-]\vee\sim A).\\&\equiv.(F[\ _+]\vee G[\ _-])\vee(A\wedge\sim A).\equiv.F[\ _+]\vee G[\ _-],\end{aligned}$$

which gives rise to $\dashv F[A_+] \wedge G[A_-]$ by the rule 3.21. Since we have $\vdash F[A_+]$ and $\vdash G[A_-]$ by Lemma 4.5, $\vdash F[A_+] \wedge G[A_-]$ is also a thesis of the Hilbert-type version of CP. But this is against Theorem 5.2 (Contradiction theorem). $F[\ _+] \vee G[\ _-]$ is, thus, provable by the tableau method for CP.

This completes the proof of Theorem 5.3, i.e. normal form theorem.

Analogously with the cases of other logical systems, normal form theorem could be employed for obtaining the equivalence between the tableau method and Hilbert-type versions of CP. It is also a corollary to this Theorem that both versions are consistent.

Nevertheless, the consistency cannot be made use of for proving Theorem 5.2, i.e. Contradiction theorem, in particular, for taking care of the basis case, if it is derivable from normal form theorem, which is, here, in turn based upon Contradiction theorem as has been shown in the above proof of normal form theorem. For avoiding such a circularity, we are advised to go back to the more traditional method for proving normal form theorem.

6 Constant Propositions

In this section, some constant propositions will be introduced both to the tableau method and Hilbert-type versions of CP. More specifically, two constant propositions to be designated as 1 and 0 will be introduced. Intuitively, they represent, respectively, constantly true and constantly false sentences.

In the Hilbert-type CP thus augmented, 1 and 0 are characterized by such axioms as,

$$\vdash 1, \ \ \vdash\sim 0, \ \ \dashv \sim 1, \ \ \dashv 0,$$

from which follows on the basis of some tautologies,

$$\vdash 1 \equiv\sim 0, \ \ \vdash\sim 1 \equiv 0.$$

Along with these truth table analogues, we also have such analogues as;

$$\begin{array}{ll}
\vdash 1 \wedge 1. \equiv 1, & \vdash 1 \wedge 0. \equiv 0, \\
\vdash 0 \wedge 1. \equiv 0, & \vdash 0 \wedge 0. \equiv 0, \\
\vdash 1 \vee 1. \equiv 1, & \vdash 1 \vee 0. \equiv 0, \\
\vdash 0 \vee 1. \equiv 0, & \vdash 0 \vee 0. \equiv 0,
\end{array}$$

which are easily derivable on the basis of these axioms and some tautologies.

The definitions of tableaux and Hintikka formulas are also requested to be modified so that constant propositions are accommodated to the proposed versions of CP. For example, the clause 4.21 of the definition of Hintikka formulas is replaced by:

4.21′ A is neither of the form $F[A_+, A_-]$ nor $F[1_+]$ nor $G[0_-]$.

The definition of closed tableaux is also changed accordingly. All the theorems thus far advanced remain valid *mutatis mutandis*. For example, given a Hintikka formula A , if we substitute 0 and 1, respectively, for propositional variables occurring in A as its positive or negative parts, we obtain a formula A' such that

$$\vdash A' \supset 0,$$

the proof of which is similar to that for Lemma 4.6.

From this immediately follows that

$$\vdash \sim 0 \supset \sim A'.$$

and, consequently, that $\sim A'$ is a thesis of the Hilbert-type as well as of the tableau method version of CP augmented with constant propositions.

Summing up,

Theorem 6.1 (Negation as failure). *If A is a formula containing only constant propositions in place of propositional variables and is not provable in the tableau method (Hilbert- type) version of CP, then $\sim$ A constitutes a thesis of the same version.*

For visualizing the situation, suppose, we are given a problem to be solved. Further, it is assumed, we have failed to solve the problem. The problem is, then, reduced to a Hintikka formula. A positive answer is forthcoming, thereupon, by negating the problem. In other words, the failure to solve the problem turns into a successful solution by negating the given problem. As shown by Clark [1], this really takes place in the implementations of the programming language Prolog, of course, in a far more complicated way.

By interpreting 1 and 0 not as constant sentences, but as the truth values truth and falsity, a model theory for CP is developed, where Hintikka formulas again play an important role. In fact, not only completeness theorem, but also many theorems such as dichotomy, contradiction and normal form theorems are forthcoming almost straightforwardly.

Before concluding this section, we wish to remark in passing that Corollary 4.9 is to the effect that every formula reduced to a Hintikka formula is axiomatically rejected, since such a formula belongs to a branch ending with the Hintikka formula.

The process of axiomatic rejection as above indicated is, however, more perspicuously integrated as a series of implications beginning Lemma 4.8, while the Hintikka formula in terun implies a disjunction of propositional variables or their negations. This is proved by induction on the length of the formulas belonging to the branch leading from the Hintikka formula down to the disjunction on the basis of Lemma 4.4, which has the effect of reducing the length of the formulas involved by way of an implication.

Summing up, we have a tautology:

$$A \supset .B_1 \vee B_2 \vee \ldots \vee B_n \quad (1 \leq n),$$

where A is the formula to be reduced and $B_1, B_2, \ldots, B_n$ are propositional variables or its negation.

The succedent of the tautology is axiomatically rejected by way of the rules 3.22 and 3.21 as shown in the proof of Lemma 4.6, and from this follows, by 3.21, that A is axiomatically rejected.

7 Conclusion

In the present paper we have developed the theory of axiomatic rejection in connection with the tableau method confining ourselves to classical propositional logic. In the future, we wish to generalise our results and cover various syllogistic systems including Lesniewski's ontology, of which the study was already started by Inoué–Ishimoto–Kobayashi [5]. The Gentzen structure inherent in syllogistic systems and programming language Prolog, it is hoped, will be explored in detail in particular connection with axiomatic rejection, the former having been initiated by Kanai [7] and the latter by Shimidzu [13].

Iwaki Junior College, Japan

References

[1] K. L. Clark. *Negation as Failure, Logic and Data Bases*, New York, 1978.

[2] V. K. Härtig. Zur Axiomatisierung der Nicht-Identitäten des Aussagenkalküls, *Zeitschrift für Math. Logik und Grundlagen der Math.*, **6**, 1960.

[3] K. J. J. Hintikka. Form and quantification theory. *Acta Philosophica Fennica*, **8**, 1955.

[4] T. Inoué. On Ishimoto's theorem in axiomatic rejection. *Philosophy of Science, Japan*, **22**, 1989. (In Japanese)

[5] T. Inoué, A. Ishimoto and M. Kobayashi. Axiomatic rejection for the propositional fragment of Lesniewski's ontology. Forthcoming.

[6] A. Ishimoto. The method of axiomatic rejection in classical propositional logic and its applications. *The Logic of Natural Language*, 1990. (In Japanese)

[7] N. Kanai. The Gentzen-type formulation of syllogistic and its completeness with respect to first-order predicate logic. *The Logic of Natural Language*, 1990. (In Japanese)

[8] M. Kobayashi and A. Ishimoto. A propositional fragment of Lesniewski's ontology and its formulation by the tableau method. *Studia Logica*, **41**, 1982.

[9] J. Łukasiewicz. *Aristotle's syllogistic from the standpoint of modern formal logic*, second edition, Oxford, 1957.

[10] K. Shütte. *Beweistheorie*, Berlin-Heidelberg, 1957.

[11] K. Shütte. *Volständige System Modaler und Intuititionistischer Logik*, Berlin-Heidelberg, 1968.

[12] K. Shütte. *Proof theory*, Berlin-Heidelberg, 1978.

[13] S. Shimidzu. The Logical Structure of Parsing, *Memoirs of Iwaki Junior College*, **13**, 1987. (In Japanese)

[14] J. Slupeci. Badan nad sylogistica Aristoteles. *Traveaux de la Société des Sciences eet des Lettres de Wroclaw*, **9**, 1948.

[15] J. Slupecki, G. Bryll and Wybraniec–Skardowska. Theory of rejected propositions. *Studia Logica*, **24**, 1971, and *Traveaux de la Société des Sciences eet des Lettres de Wroclaw*, **30**, 1972.

[16] R. M. Symullyan. *First-Order Logic*, Berlin-Heidelberg-New York, 1968.

[17] W. Staszek. A certain interpretation of the theory of rejected propositions. *Studia Logica*, **30**, 1972.

[18] J. Wolenski. *Logic and Philosophy in the Lvov–Warsaw School*. Kluwer, Dordrecht, 1989. (Polish edition, 1985.)

Selected Bibliography of Vladimir Aleksandrovich Smirnov's Works

[1] V. A. Smirnov. On the theory of categorical syllogism. *Filosofskie nauki*, **3**, 1959. (In Russian)

[2] V. A. Smirnov. Role of symbolisation and formalisation in cognitive processes. *Proc University of Tomsk, Tomsk*, 1960. (In Russian)

[3] V. A. Smirnov. Genetic method of scientific theory constructing. In *Philosophical Problems of Contemporary Formal Logic*, Nauka, Moscow, 1962. (In Russian)

[4] V. A. Smirnov. Some conclusions from the comparison of Markov's normal algorithms with Lyapunov's algorithm schemata. *Proc University of Tomsk, tomsk*, 1962. (In Russian)

[5] V. A. Smirnov. Logical ideas of N. A. Vasiliev. *Essays in History of Logic in Russia*, Moscow State University,Moscow, 1962. (In Russian)

[6] V. A. Smirnov. Notes on syllogistics and general deduction theory. *Proglems of Logic*, Nauka, Moscow, 1963. (In Russian)

[7] V. A. Smirnov. Algorithms and algorithm logical schemata. *Problems of Logic*, Nauka, Moscow, 1963. (In Russian)

[8] V. A. Smirnov. On degnity and mistakes of one philosophical conception (critical notes on the theory of language constructions of R. Carnap). *Philosophy of Marksism and Neopositivism*, Moscow State University, 1963.

[9] V. A. Smirnov. Levels of knowledge and stages in the process of knowledge. *Problems of the Logic of Scientific Knowledge*, Nauka, Moscow, 1964. (In Russian)

[10] V. A. Smirnov, I. S. Narsky and P. V. Kopnin. Serious shortages of the book on materialistic dialectics. *Voprosy Filosofii*, **4**, 1964. (In Russian)

[11] V. A. Smirnov. Models of language and models of world. *Problems of the Logic of Scientific Knowledge*, Nauka, Moscow, 1965. (In Russian)

[12] V. A. Smirnov. Review on the book of N. I. Styashkin 'Formationof Ideas of Mathematical Logic'. *Voprosy Filosofii*, 1966. (In Russian)

[13] V. A. Smirnov. Embedding of the syllogistics into predicate calculus. *Logical Semantics and Modal Logic*, Nauka, Moscow, 1967. (In Russian)

[14] V. A. Smirnov. Wissensebenen und Etappen des Erkenntnis. *Studien zur logik der Wissenschaftlichen Erkenntniss*, 1967. (In German)

[15] V. A. Smirnov. Natural deduction and transformation analysis. *Proc 14th Int. Congress of Philos, Vienna*, 1968. (In Russian)

[16] V. A. Smirnov. Thrid International Congress on Logic, Methodology and Philosophy of Science. Scientific Technical Information, ser. 2, 1968. (In Russian)

[17] V. A. Smirnov. So-called paradoxes of material implication and logical systems with notion of strong deduction. *Investigation on Logical Systems*, Nauka, Moscow, 1970. (In Russian)

[18] V. A. Smirnov. Syllogistics without law of excluded middle. *Investigation on Logical Systems*, Nauka, Moscow, 1970. (In Russian)

[19] V. A. Smirnov. Levels of knowledge and stages in process of knowledge. *Porblems of the Logic of Scientific Knowledge*, Reidel,Dordrecht, 1970.

[20] V. A. Smirnov. Eliminatin des termes dans la logique intuitioniste. *Revue Intern. de philosophie*, **98**, 1971.

[21] V. A. Smirnov. On decidability of decision problem for sequential calculus of preducates without contractions. *Proc IV Int. Congress on Logic, Mehtodology and Philosophy of Science, Bucharest*, 1971.

[22] V. A. Smirnov. Syllogistics in sequential form without propositinal calculus. *Filosofskie Nauki*, **3**, 1972. (In Russian)

[23] V. A. Smirnov. *Formal Deduction and Logical Calculi*, Nauka, Moscow, 1972. (In Russian)

[24] V. A. Smirnov. Review on the book of O. F. Serebryannikov 'Heuristic Principles and Logical Calculi'. *Voprosy Filisofii*, **2**, 1972. (In Russian)

[25] V. A. Smirnov. An absolute first-order predicate calculus. *Bull of the Sec of Logic, PAN*, 1973.

[26] V. A. Smirnov. On relations between symbolic logic and philosophy. *Philosophy in the Contemporary World. Philosophy and Logic*, Nauka, Moscow, 1974. (In Russian)

[27] V. A. Smirnov. On definability of the predicates introduced by two-side recuction propositions. *Theory of Logical Inference, Part II*, Moscow, 1974. (In Russian)

[28] V. A. Smirnov. Sequent variant of logical systems with strong and relevant implication. *Philosophy in Contemporary World. Philosophy and Logic*, Nauka, Moscow, 1974. (In Russian)

[29] V. A. Smirnov. Introductory article to the book of Ingols 'Introduction to Indian Logic Nvja–Njaja'. Nauka, Moscow, 1974. (In Russian)

[30] V. A. Smirnov. A new form of the deduction theorem for P, E and R. *Proc 5th Int Congress of Logic, Methodlogy and Philosphy of Science, London, Canada*, 1975.

[31] V. A. Smirnov, I. A. Akchurin, U. L. Ershov and V. N. Sadovsky. Fifth Internatinal Congress on Logic, Methodology and Philosophy of Science. *Voprosy Filisofii*, **2**, 1975. (In Russian)

[32] V. A. Smirnov. Definability and Identifiability. Certain Problems and Hypotheses. *Basic Problems in Methodology and Lingusitics*, Dordrecht, Boston, 1977.

[33] V. A. Smirnov. Formal deduction, deduction theorems and theories of implication *Logical Deduction*, Nauka, Moscow, 1978. (In Russian)

[34] V. A. Smirnov. Theory of Quantification and ε-calculi. *Essays on Mathematical and Philosophical Logic*, Dordrecht, 1979.

[35] V. A. Smirnov. Logical systems with modal temporal operators. *Proc 2nd Finnish–Soviet Colloquium*, Moscow, 1979. (In Russian)

[36] V. A. Smirnov. Logical systems with modal temporal operators. *Proc VI Int Congress of Logic, Methodology and Philosophy of Science, Hannover*, 1979.

[37] V. A. Smirnov,V. A. Boxharov, E. K. Vojshvillo and A. G. Dragalin. Some problems of development of logic. *Voprosy Filisofii*, **6**, 1979. (In Russian)

[38] V. A. Smirnov. Adequate translation of syllogistic assertions into predicate calculus. *Actual Problems of Logic and Methodology*, Naukova dumka, Kijev, 1980. (In Russian)

[39] V. A. Smirnov. Logical rendszerek modalis idöperatorokkal. *A filosofia idöczerü kerdesei*, **42**, 1980.

[40] V. A. Smirnov and V. N. Sadovsky. J. Hintikka and the development of epistemological investigatins int he second part of the twentieth century. *J. Hintikka, Logical and Epistemological Investigations*, Progress, Moscow, 1980. (In Russian)

[41] V. A. Smirnov. Modern sematnical investigations in modal and intensional logics. *Semantics of Modal and Intensional Logics*, Progress, Moscow, 1981. (In Russian)

[42] V. A. Smirnov. Embedding of the elementary ontology of St. Lesneiwski into monadic second-order preicate calculus. *Modal and relevant logics. Proc of the Sci. Seminar of Institute of Philosophy of AS USSR*, Moscow, 1982. (In Russian)

[43] V. A. Smirnov. The definition of modal operators by means of tense operators. *Acta Philosophica Fennica*, **35**, 1982.

[44] V. A. Smirnov. Logical methods of comparison of theories. *Voprosy Filosofii*, **6**, 1983. (In Russian)

[45] V. A. Smirnov. Tense lgoics with nonstandard interconnections between past and future. *Proc VII Int. Cong. Logic, Methodology and Philosophy of Science, Vol 2, Salzburg*, 1983.

[46] V. A. Smirnov. Logische Bezungen Zwischen Wissenschaftlichen Theorien. *Logic, Methodologie und Wissenschaft*, Moscow, 1983. (In German)

[47] V. A. Smirnov. Sequent formulation of Dummett's logic. *Semiotical apsects of formalisation of mental activity*, Moscow, 1983. (In Russian)

[48] V. A. Smirnov. Topological interpreationof modal syllogistics. *Proc Symp Aristotle's Logic*, Metsniereba, Tbilisy, 1983. (In Russian)

[49] V. A. Smirnov. Embedding of the positive syllogistics into monadic calculus of predicates. *Logical investigation. Proc of Sci.Seminar of Institute of Philosophy of AS USSR*, Moscow, 1983. (In Russian)

[50] V. A. Smirnov. Definitional equivalence of extended syllogistics C2D to Boolean algebra. *Logical investigation. Proc of Sci.Seminar of Institute of Philosophy of AS USSR*, Moscow, 1983. (In Russian)

[51] V. A. Smirnov. Embeddingof the elemntary ontology of St. Lesniewski into monadic second-order calculus of predicates. *Studia Logica*, **XLII**, 1983.

[52] V. A. Smirnov. Loic, foundations of mathematics and linguistics. *Voprosy Filosofii*, **1**, 1984. (In Russian)

[53] V. A. Smirnov. On a system of parraconsistent logic. *Many-valued, relevant and paraconsistent logics. Proc of Sci.Seminar of Institute of Philosophy of AS USSR*, Moscow, 1984. (In Russian)

[54] V. A. Smirnov. The definition of modal operators by means of tense operators. *Modal and Intensional Logics and their Application to Methodology of Science*, Nauka, Moscow, 1984. (In Russian)

[55] V. A. Smirnov. Logical systems with modal temporal operators. *Modal and Intensional Logics and their Application to Methodology of Science*, Nauka, Moscow, 1984. (In Russian)

[56] V. A. Smirnov. Recosntruction of modal syllogistics. *Intensional Logics and Structure of Scientific Theories*, Metsniereba, Tbilisy, 1985. (In Russian)

[57] V. A. Smirnov. Symbolic logic and cognitive science. *Philosophy and Sociology of Science and Technique*, Nauka, Moscow, 1985. (In Russian)

[58] V. A. Smirnov. Logical relations between theories. *Synthese*, **66**, 1986.

[59] V. A. Smirnov. G. H. von Wright's contributin to logic and philosophy of science. Preface to G. H. von Wright *Logical and Philosophical Investigations. Selected Works*, Progress, Moscow, 1986. (In Russian)

[60] V. A. Smirnov. Assertion and predication. Propositional logic and logic of events. *Nonstandard Semantics of Nonclassical Logics*, Nauka, Moscow, 1986. (In Russian)

[61] V. A. Smirnov. Modality de re and Vasiliev's imaginary logics. *Logique et Analyse*, **114**, Lenwain, 1986.

[62] V. A. Smirnov. Creation, discovery and logical methods of proof search. *The Nature of Scientific Discovery*, Nauka, Moscow, 1986. (In Russian)

[63] V. A. Smirnov. Logical analysis of scientifc theories and relations between them. *Logic of Scientific Knowledge*, Nauka, Moscow, 1987. (In Russian)

[64] V. A. Smirnov. *Logical Methods of Analysis of Scientific Knowledge*, Nauka, Moscow, 1987. (In Russian)

[65] V. A. Smirnov. Logical ideas of N. A. Vasiliev and modern logic. *Abstracts of VIII Int. Cong. Logic, Methodology and Philosophy of Science*, Moscow, 1987.

[66] V. A. Smirnov. Assertion and predication. Combined calculus of propositions and sitations. *Abstracts of VIII Int. Cong. Loigc, Methodology and Philosophy of Science*, Moscow, 1987.

[67] V. A. Smirnov. Epistemology and symbolic logic. Science as a subject of study. *Social Sciences Today*, Moscow, 1987.

[68] V. A. Smirnov. Strict embedding of theelementary ontology intothe monadic second-order calculus of predicates admitting the empty individual domain. *Studia Logica*, **XLVI**, 1987.

[69] V. A. Smirnov. Actual philosophical investigation. *Philosphy and Sociology of Science and Technique*, Nauka,Moscow, 1987. (In Russian)

[70] V. A. Smirnov. Problems of investigation oflanguage and analytic philosophy. *Voprosy Filosofii*, **8**, 1988. (In Russian)

[71] V. A. Smirnov. Intensional logics and logical structure of theory. *Porc IV Finnish–Soviet Symposium on Logic, Tbilisy*, 1988. (In Russian)

[72] V. A. Smirnov. Assertion and predication. Combined calculus of sentences and events. *Intensional Logic, History of Philosophy and Methodology. To Imre Ruzsa on the occassion of his 65th Birthday*, Budapest, 1988.

[73] V. A. Smirnov and S. N. Artjomov. Logic and its perspectives. *Voprosy FIlosofii*, **6**, 1988. (In Russian)

[74] V. A. Smirnov. Combining of the calculi of sentences and events with von Wright's logic of truth. *Investigations on Nonclassical Logics*, Nauka, Moscow, 1989. (In Russian)

[75] V. A. Smirnov. Assertion and predication. Combined calculus of sentences and events. *SYntactic and Semantic Investigations of Nonextentional Logis*, Moscow,1989. (In Russian)

[76] V. A. Smirnov. Logical ideas of N. A. Vasiliev and modern logic. Preface to N. A. Vasiliev's *The Imagine Logic*, Nauka, Moscow, 1989. (In Russian)

[77] V. A. Smirnov. Creation, discovery andlogical methods of proofs. *Filosofskaja i sotsiologichekaja mjisl*, **3**, Kijev, 1989. (In Russian)

[78] V. A. Smirnov. Logical ideas of N. A. Vasiliev and modern logic. *Logic, Methodology and Philosphy of Science*, **VIII**, Amsterdam, New York, 1989.

[79] V. A. Smirnov. E. Kant and modern logic. *Kantovskij sbornik*, **14**, Kaliningrad, 1989.

[80] V. A. Smirnov. Logical analysis of Kant's theory of justice and moral. *Kantovskij sbornik*, **15**, Kaliningrad, 1990. V. A. Smirnov.

[81] V. A. Smirnov. Proof search. *Logic and Computer*, Nauka, Moscow, 1990.

[82] V. A. Smirnov. On multidimensional logics and their relations with many valued logics. *Proc X Conf. Logic, Methodology and Philosophy of Science, Minsk*, 1990.

[83] V. A. Smirnov. Problems of truth in logical semantics. *Theory of Cognition, vol. 2*, Mysl', Moscow, 1991. In Russian.

[84] V. A. Smirnov. Multidimensional logics. *Porc IX International Cong. Logic, Method. and Philos. of Science*, Uppsala, 1991

[85] V. A. Smirnov. M. K. Mamardashvili: Philosophy of consciousness. *Kommunist*, **8**, 1991. In Russian.

[86] V. A. Smirnov. On memory of Professor O. F. Serebriannikov. *Voprosy Filosofii*, **10**, 1992. In Russian.

[87] V. A. Smirnov. A. I. Rakitov: Philosophy of computer revolution. *Voprosy Filosofii*, **1**, 1993. In Russian.

[88] V. A. Smirnov. Double aglebras and symmetric logics. *Logical Investigations, vol. 1*, Nauka, Moscow, 1993. In Russian.

[89] V. A. Smirnov. Cut elimination in e-calculus. *Book of Abstracts, 1, XIX World Congress of Philosophy*, Moscow, 1993.

[90] V. A. Smirnov. Definitional equivalence of Lesniewski's ontology and generalized syllogistics with axiom of atomicity. *Logical Investigations, vol. 2*, Nauka, Moscow, 1993. In Russian.

[91] V. A. Smirnov. Multidimensional logics. *Logical Investigations, vol. 2*, Nauka, Moscow, 1993. In Russian.

[92] V. A. Smirnov. Definitional equivalence of elementary ontology and generalized syllogistics of Occamian type. *Preprint 93-03*, Institute for Logic, Cognitive Sciences and Development of Personality, Moscow, 1993.

[93] V. A. Smirnov and A. Melentjev. *Logic and Clinical Diagnostics: Theoretical Basis*. Nauka, Moscow, 1994. In Russian.

[94] V. A. Smirnov. Definitional equivalence of the syllogistic systems. *Proceedings of the Research Logical Seminar of Institute of Philosophy, Russian Academy of Sciences*, Moscow, 1994. In Russian.

[95] V. A. Smirnov. Reminiscences of teacher. *Voprosy Filosofii*, **1**, 1995. In Russian.

[96] V. A. Smirnov. Popper is right: dialectical logic is impossible. *Voprosy Filosofii*, **1**, 1995. In Russian.

[97] V. A. Smirnov. Proof search in natural intuitionisitc predicate calculus. *XI International Conf. Logic, Methodology and Philosophy of Science, II*, Moscow-Obninsk, 1995. In Russian.